Invisible Robots in the Quiet of the Night

How AI and Automation Will Restructure the Workforce

Craig Le Clair

Forrester

TABLE OF CONTENTS

1 The Forces Of Automation Are Upon Us 1

2 Control Passes From Humans To Machines 11

3 A New Formula For Scale 21

4 Convergence Brings The Physical And Digital Worlds Together 31

5 Black Mirror Forecasts And The Perils Of The S-Curve...........41

6 Automation Deficits Are Inevitable 47

7 Automation Dividends Don't Fill The Employment Gap.........63

8 The Murky Middle Challenges Job Transformation.................71

9 The Talent Economy 81

10 It's Different This Time — It Really Is 91

Acknowledgements...105

About The Author..107

Methodology...109

Endnotes..111

Index..123

1 THE FORCES OF AUTOMATION ARE UPON US

The book you're reading is full of data from Forrester and others. But it's also rich with stories of people whom automation has affected both positively and negatively — the machinist, the ride-sharing gig worker, the insurance underwriter, the displaced coal miner in Appalachia, the factory worker, and the security guard. The book comes complete with a four-part model of the future workplace and the 12 work personas that will occupy it.

If you're a business or individual with a traditional view of work, you risk being run over by the future. The old view of work belongs to a time of "while you were out" slips, pocket protectors, and voicemail. Businesses and workers clinging to the remnants of this now mythical workplace will find that grip getting weaker.

Opinions on the future of work are divided. The historians, as we call them, believe that normal work patterns will absorb AI, as with previous automations. It'll be business as usual. Among the competing views, the historians have the longest and richest data set. They argue that automation has been on a never-ending upward arc to reduce human toil: Sticks and stones improved the odds for hunting game, while the wheel advanced transport. Each new generation marveled at the progress, but in every case, the initial disruption created more jobs than it destroyed. Workers just shifted to different jobs.

Robot enthusiasts make up another camp. They see the robots lurking around every corner and can't wait. Robots will free workers from mind-numbing tasks that they believe are beneath them, anyway. Entering data into spreadsheets or answering mindless questions from customers just isn't meaningful work. Robots will do it better. The tech industry and businesses that see opportunity in automation tend to support this view.

The most dramatic views, by far, are held by what we call the dystopians. Bolstered by impressive thinkers like Bill Gates, Elon Musk, and Stephen Hawkins, they believe we're in the front-row seat of a post-civilization horror movie. Robots will take all our jobs and then, as in a zombie apocalypse, head to the streets. The dystopians believe that the forces of automation will bring us a society of haves and have-nots, a world where the rich exploit the poor from luxury condominiums. Meanwhile, the masses below eke out dreary lives and struggle against declining wages and anxiety about making frayed ends meet. It's the world pictured in the 1982 movie *Blade Runner*, which featured a crumbling infrastructure and collapsing social order, a world of decadence and decay.

But our view isn't as black or white as these. The future of work will evolve in a more gradual manner. Over the next 10 years, millions of people will depart the "golden age of work" and head to one of four places: They'll become automation deficits; find new work as automation dividends; enter the talent economy; or start work in a job transformed by machines. The transition will be bumpy — digital outcasts, new forms of digital anxiety, and pockets of social unrest will increase. The pace is gradual for two reasons: Firstly, our institutions, both private and public; labor unions; regulations; and culture will battle with the advance of technology. Secondly, AI is a general-purpose technology (GPT), just as electricity, phones, and the internet are, and like previous GPTs, it will take some time to get going.

The first part of the book describes the forces of automation (FOA) will that make this happen. Note that we use the terms AI and automation interchangeably throughout. Automation is the broader term and includes technologies that range from a simple calculation in a spreadsheet to the most advanced deep-learning AI algorithms. AI would be a big circle, but it would still be inside a much bigger one called automation.

Collectively, these automations have three major effects on the workforce: control, scale, and convergence (see Figure 1-1). We devote a chapter to each. Chapter 2 examines the shift of control from humans to machines. Previous automations made work easier, but they left humans in charge. AI will move control rapidly to the machines themselves. This starts to reverse a 2012-to-2018 shift of control from companies to customers who use social and mobile technologies to access product reviews, comparison sites, and apps that could bypass traditional companies.

Chapter 3 explores a new formula for scale and its implications. Common application platforms, cloud infrastructure, and mobile networks, fueled by machine learning, will allow business to scale at unprecedented levels but will require fewer workers.

Chapter 4 describes the third force of automation: convergence. This automation brings the physical and digital worlds together and alters many jobs. Physical products, equipment, buildings, and factories make up the physical world. It has a long history, and by comparison, the digital world is new. It produces data from bank accounts, social media, and internet of things (IoT)-generated sensor data. When these two worlds come together, the result is convergence.

Figure 1-1 The Forces Of Automation Will Transform Work

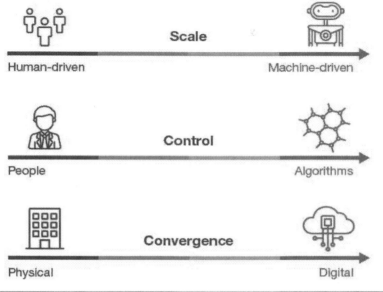

Automation And Jobs Is An Uncomfortable Subject

The impact of AI and automation on work is a difficult conversation, one that runs afoul of business, political, and media agendas. Conservatives speak to the unemployed at factory gates, coal mines, and docks, blaming the workers' struggles on soft borders, unfair trade practices, and excessive regulation. Liberals point to insufficient investment in education and resistance to minimum-wage legislation. Business points to AI's potential to free workers from mundane tasks to be more human, but behind closed doors, these same companies draft business cases to reduce headcount.

And confusion reigns. AI applications can involve 15 or more different capabilities, all with variations that shift with academic and business innovation. Headlines keep the megaphone horizontal with daily reports of AI breakthroughs. The tech industry is perhaps the loudest voice. These players know that the next great pool of wealth will derive from AI, and they all want their share. As we'll see, this makes it difficult to separate the art of the possible from reality.

What no camp tells the worker is this: Automation is being developed around the world to replace you or significantly restructure your work. Governments and industry are investing billions in AI startups and megacloud platforms that, if successful, will replace millions of workers and restructure most jobs. The intent isn't evil. It's just Silicon Valley capitalism at work, the latest stage of innovation that has been progressing for centuries. The success of AI investment just happens to require worker displacement or radical job transformation. One can't happen without the other.

These commercial and political agendas, as well as general confusion, are challenging a balanced view. But agendas aside, we can all agree on one thing: Right now, workers are concerned over a future with few jobs. And can we blame them? Creepy robot images are everywhere. The 2019 Super Bowl commercials featured robots or smart devices in no fewer than 10 ads. One of them really hit home: It showed two men in conversation. One points out, "All I'm saying is, in five years, robots will be able to do your job." The speaker says this sitting at a baseball game with a robot eating a hot dog behind him. OK, that's a commercial. But AI is now all around us. Pandora learns our taste in music. Amazon tells us the weather at the kitchen table and may someday use our conversations to target ads.[1] Netflix uses machine learning to improve our movie lists.

But consumer-directed applications of machine learning are an early form of "AI lite" and don't materially affect workers. For maximum job effects, mature companies must make AI work in the business of work, and so far, that hasn't happened. The laggard nature of business adoption is a familiar story. Mobile innovation took forever; it's now 10 years since Apple introduced the iPhone, and many firms still don't have easy and reliable mobile support. Mobile is still more of a bolt-on to fragile legacy systems than an integrated customer experience. Adoption of AI is likely to follow a similar pattern. The point is simple but important: Mature companies employ most workers today, and they're behind in deploying AI. Their journeys are just starting. The Future of work will depend on how rapidly and completely mainstream business adopts AI. This is our focus.

Machine Learning And Robotic Process Automation (RPA) Underpin The FOA

The first generation of disruptive digital platforms, such as Airbnb and Uber, connected users, who couldn't previously reach other, to form new markets. Elegant discovery, matching, and location algorithms bypassed traditional companies. From a job perspective, and compared with where we're now heading, they were benign. Sure, they turbocharged the gig economy, creating a new generation of on-demand workers, and altered occupations like taxi drivers and small hotel operators. But these early and disruptive digital platforms didn't affect traditional employment as much as the next phase of automation will. Here's why: Machine learning is powering that next automation wave.

You don't need a deep understanding of machine learning to understand the future of work. The important thing is not how the algorithms differ but rather what they have in common: the power to learn. Most machine learning today takes in data and runs it through static calculations. But machine learning is evolving to use incoming data to reconfigure and improve models without human intervention. There's no master program or spreadsheet tracking what you like to watch on Netflix or what you like to buy at Amazon. Algorithms are automatically updating your personal history to sharpen your categorization for various commercial purposes, some that you know about and some that you don't. This power to learn is the core of automation's disruptive potential.

There's a second major innovation that not nearly enough people are paying attention to — and that's the software robots that will take the most jobs. Software running in obscure data centers that no one will ever see will replace or transform the jobs of cubicle workers, coordinators, and even knowledge workers. These are the invisible robots that RPA platforms are building. Put simply, these platforms build bots that mimic what a human does on a computer. Repetitive tasks, such as posting data to a financial application, are easily programmed into the bot. And unlike many machine learning projects that may need data scientists, RPA is simple to deploy. It's now an explosive enterprise software market. The valuation of the top three RPA software companies, which almost no one had heard about two years ago, is now in the range of $11.3 billion.[2] RPA doesn't require the "moonshot" approach that many AI projects do and is quietly piling up automation deficits in our cubicle worker and coordinator personas. Chatbots are yet another category of software robot that are making inroads in customer service. Indeed, a primary argument you'll see throughout this book is that we're too focused on physical robots — while the invisible ones are preparing to wreak havoc.

People, Not Machines, Will Write The Next Chapter

As science fiction writer William Gibson once said, "The future is already here — it's just not evenly distributed."[3] This means you can find it, but it takes some looking. Changes aren't obvious or everywhere, but you'll see them in small segments and examples. For example, if you looked carefully in 2001, you could find a small set of individuals addicted to their BlackBerry phones. They had a bit of swagger. They disrupted meetings to check emails and annoyed their spouses with lack of attention. People even called the now-crude devices "crackberries." Think about it. This was six years before Apple introduced the iPhone and 10 years before mobility exploded. Now, most of us fall asleep with our devices nearby. In other words, if we had looked for them, we might have seen the signs of device addiction very early.

Likewise, if we look hard enough, we can see the future of work by observing people who interact today with intelligent machines. We've interviewed scores of them. We've learned how workers will take to machines that are learning faster than they can. From these discussions, we've built a framework that addresses a post-automation view of work, new categories of workers, and new work patterns that have humans and machines working more closely together.

Automation is a tale of winners and losers. It will work for many: It helped Roxy Prima, who's built an art business using Instagram; and Nissa Scott, a worker from Amazon who takes advantage of robots to do traditionally male heavy lifting; and Prasad Akella, one of the digital elites, who's now on his third startup on factory automation.

It's easy to be blinded by the potential of intelligent machines that beat humans in complex games or guide automobiles effortlessly. But automation must work for people, and for many, it doesn't fit into their existing work patterns, psychology, or economic reality. For example, people like Enzo, a machinist who had to quit because he couldn't program his machine reliably; or Julie, an accounts payable clerk whose job was taken by a software robot. Or the insurance underwriter who has to train a machine learning algorithm to make loan decisions; or the airport ticket agent brought to tears as she can no longer make decisions that help passengers; or Sally, a customer service agent who doesn't appreciate chatbots; or Jill, a bank teller closely monitored by AI systems and struggling to make $15 dollars an hour. And then there's Nick Mullins, a fifth-generation coal miner who tried his hand at contact center work in Appalachia. Many will pursue alternate work styles, like David White, a mission-based worker who now takes care of foster kids, or Rich Lane, who tells of his many scars from the gig economy.

These are stories about how people solve thorny problems when their jobs are gone or when automation has altered them beyond recognition. It's about how they seize automation dividends, navigate the automation deficits, drift to alternate work styles, or pursue life without work. And it's about how their examples can inspire or guide the rest of us.

The Business Viewpoint On Automation Is Limited

Enterprises today invest in AI to reduce costs or as a defensive measure to stay even with or get ahead of their peers. But automation can be much more than that. It can lift employees to new levels of commitment, energy, and productivity; allow a more human face to a brand; and take customers to new experiences.

But for this to happen, enterprises must embrace tomorrow's workforce — and that will be a mix of full-time employees and talent with no formal ties to a company at all. Workers will move from role to role and across organizational boundaries more freely than ever before. Enterprises must replace the cultures and systems they developed with an "owned talent" mentality. The gig economy that exploded to support the sharing economy will become something more — a talent economy — and will permanently change how we all work.

We See Four Primary Options And 12 Work Personas In The Future Of Work

When businesses grab hold of automation, it will cannibalize some jobs, create others, and transform most. But to understand these forces, we need a working model of the future, one based on where today's human workers end up. We also need a new way to classify workers that helps enterprises plan and helps employees develop their careers. We provide both.

Broadly, there are only four work options (see Figure 1-2). The traditional work environment, under pressure by automation, will push today's 158 million US workers into one of these four categories: 29% will become automation deficits; 80% will have their jobs transformed by automation in a significant way; and 30% will depart for the talent economy. On the plus side, 13% new jobs will be created as automation dividends.[4]

Figure 1-2 The Four Options For The Future Of Work

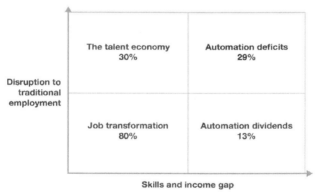

Source: Forrester forecasts

How automation will transform the workplace is important. How many deficits will there be, and which jobs will be most affected and when? What will the automation dividends look like? How will the gig economy transition to something more formal? And how will automation transform the jobs that remain? But to make the model actionable, we need a view from the worker's perspective. To organize that view, Forrester created 12 new work personas that capture more than 800 current occupations tracked by the US Bureau of Labor Statistics (BLS). We define the personas in today's workforce and break them into knowledge, frontline, and administrative categories (see

Figure 1-3). We also define the personas that will emerge in the future (see Figure 1-4). We'll explore all of these in detail in later chapters.

Figure 1-3 Today's Traditional Personas Do Knowledge, Frontline, And Administrative Work

Traditional economy workers		Profile/definition	Examples
Knowledge work	Cross-domain knowledge workers	Workers determine tasks, ideas, priorities, artistic contributions, and goals, with insights and decisions they draw from a number of knowledge domains.	Emergency room physician
	Single-domain knowledge workers	Workers determine some tasks, priorities, and goals and draw from a single knowledge domain for insights and decisions.	Actuarial
	Function-specific knowledge workers	Structured and semistructured tasks, e.g., compiling, categorizing, calculating, auditing, or verifying information, are organized around a discrete function.	Insurance underwriter
Frontline work	Physical workers	Workers perform physical activities that require arms, legs, and moving the body, such as climbing; lifting; walking; stooping; and scaling ladders, scaffolds, or poles.	Factory worker
	Human-touch workers	Tasks include personal assistance; medical attention; and emotional support to coworkers, customers, or patients. Physical contact often combines with oral communication.	Massage therapist
	Location-based workers	Workers depend on a unique physical environment, e.g., a retail store or a secured office building. Physical environments define their jobs.	Retail store clerk
Administrative work	Coordinators	Tasks include administrative, staffing, monitoring, or controlling activities, e.g., for fleets or money spending, and providing information to supervisors, coworkers, or subordinates.	Fleet manager
	Cubicle workers	Workers perform repetitive and structured tasks in back-office and front-office positions, including workers in low-cost economies who generally perform contact center (phone) or BPO (data entry) functions.	Accounts payable administrator

Source: Forrester Research, Inc. Unauthorized reproduction, citation, or distribution prohibited.

Figure 1-4 Emerging Personas Will Reshape The Future Of Work

Emerging personas		Profile/definition	Examples
Emerging models of work	Mission-based workers	Workers believe that job satisfaction, work-life harmony, and alignment to their values and needs are important work considerations.	Yoga instructor
	Teachers/explainers	These workers know methods for curriculum design, teaching, and instruction for individuals/groups or can present machine logic and decisions.	Knowledge-based curator
	Digital elites	Enterprise architects, software development pros, and ML algorithm specialists use computers and data modeling to process information.	Data scientist
	Digital outcasts	These workers are unable to work effectively with machines or transition due to skills, attitudes, and ambitions.	Finance and accounting clerk

Who Should Read This Book

This book is an unapologetic wake-up call for governments, educators, and policy enthusiasts. If your interest is government policy, you'll benefit from a strong future-of-work framework, a realistic and practical look at the underlying technology, and a view from the worker's perspective. The rear-view mirror is the best place to view employment policy based on the old pre-automation economy. Governments need to reimagine the workplace. For example, their estimates of job growth don't account for automation trends or the emergence of the talent economy.

If your main interest is business, you can understand how to apply the FOA in positive ways to enhance customer experience and avoid the negative tendencies that can result from control, scale, and convergence factors. You need this viewpoint to recruit and manage emerging work personas like mission-based workers, who aspire to more than just regular paychecks. You can use the framework to embrace the talent economy or tomorrow's change management challenges, which will stem from human and machine collaboration. Businesses can begin to prepare for next-generation governance to control the increasingly opaque environment inherent to an algorithmic-driven business based on machine learning.

The book can help prepare individuals for 21st-century employment. For example, if you're preparing for a career, focus less on mastering a body of knowledge that will become part of a machine knowledge base and more on how to construct the decision framework that machines will use. And finally, if your main interest is computer science or engineering, reflect on the implications of your labors to understand the broader context of your intellectual pursuits.

If your main interest is the future of work, you'll get a balanced view of how and when automation will affect workers. You'll understand automation deficits, why work isn't working for many, and what dividends will result when workers adapt to the FOA.

2 CONTROL PASSES FROM HUMANS TO MACHINES

You arrive early at the airport and see an opportunity to get home in time to put your daughter to bed. The airline agent wants to put you on that early plane. She smiles, listens to your story, and tells you she has kids herself. The plane has plenty of room. Your later flight might even be oversold — but the rules in the system won't allow you to take the earlier flight without paying an exorbitant change fee. The agent wants to help, but she can't go around the rules in the system. She's diminished by the loss of authority, and you, the customer, are powerless. This ticket agent scenario is going to become more common — and also more severe. Machines will make more decisions that will make humans feel less important. The effect of AI on worker psychology is a growing topic, but that's just one concern. This chapter looks at four problems that all stem from control moving to machines: 1) psychology; 2) unpredictability; 3) data privacy; and 4) wage deterioration.

Control Shifts From Workers To Machines — And Machines Will Give Us An Attitude

Throughout most of history, humans made decisions. In the 1950s, programmers began to put rules into computers; for example, to issue an airline ticket. With advancement in software environments, coding rules into computers became easier over time, but here's what's different now: AI can connect hundreds or thousands of variables to make crisper decisions. Humans can think only about a small number of inputs at a time. And new data can automatically improve the variables that AI uses. As a result, decisions will be less costly and better. AI will move from job to job and reduce or eliminate decision-making responsibility from humans.

But are humans ready for the shift? Take self-driving cars. Many argue that they're safer now. Google's G-car, a project of Waymo, Google's sister company under the Alphabet Umbrella, has logged 10 million miles without incident.[5] AI-based self-driving cars could potentially eliminate 90% of human errors, and that's even before 5G networks could possibly improve performance.[6] But there will be literal speed bumps. Tesla's autopilot had its first lethal crash in May 2016.[7] The algorithm confused the white side of an attached trailer with the sky, and the car collided with it. It didn't help that the driver may have been watching a Harry Potter movie at the time. Ted Schadler of Forrester believes that self-driving car forecasts are overstated:

> Humans are nowhere near ready, nor are the rules needed to govern their widespread use, although the technology is advancing rapidly. The problem is entropy, the unexpected. We can't possibly know all the consequences. We're introducing an alien driving machine into an established environment, one where there are already 30,000 US deaths a year, none of them planned.
>
> Basically, Ian Malcolm (Jeff Goldblum) in the original Jurassic Park movie had it right. He predicted the downfall of the dinosaur park based on chaos theory, the idea that small changes in complex systems can have big, unpredictable effects. The park was an accident waiting to happen.
>
> We need more experience and rules to govern self-driving, like HOV lanes that might be dedicated for their use. Things that work in small scale are usually just fine. It's when you scale them up that the unpredictable can occur. This is why pharmaceutical companies have clinical trials that last for years.

In short, few would argue that driverless vehicles don't represent the future. Someday, our descendants will giggle and feel aghast that their great-grandparents drove their own cars. How mad could they be? They'll say, "They had visual distractions, texting, drugs, alcohol, and declining hand-eye coordination." But today, automated technology is further along than our attitudes. And even if humans were ready, our legal, insurance, and supporting systems aren't. For example, theoretically, self-driving cars might cut fatalities from 30,000 to 1,000. But would we use self-driving cars if consumers filed 1,000 lawsuits against the creators of the self-driving algorithms?

Giving up driving requires a major shift in psychology. If you look closely, you can even see control anxiety in elevators among your fellow passengers. In many cities, hotels and office buildings now use "destination control" elevators. You simply enter your floor and receive directions to the best elevator. Travel time is reduced, as the elevator makes fewer stops. But the elevator has no controls. Passengers step in and look for buttons that don't exist. This loss of control provokes anxiety, especially for people with various forms of phobias. Elevators aren't their favorite places to begin with — they surely don't want to get stuck in one. Some may even miss elevator operators: To be a good elevator operator, you had to like people.[8] You got to know your regular riders, and they got to know you. The human elevator operator gave us a human connection and a sense of trust that many now miss.

Here's the point: Automation is taking us further and further away from many human experiences. The more we depend on technology, the less we need other humans, or even act like humans ourselves. Interacting more with machines than with humans will change the way we think and act at work and home. And we need to understand the impact now.

Thelonius Monk And "Toronto Moments"

Unpredictability will soon become a predictable condition. AI systems differ from preceding automations. In automations past, the decision logic was transparent and readily understood. Music lovers might think of the difference by comparing how players approach classical and jazz piano. The classical pianist works from a fully notated set of music. Every note to be played is defined on the bass and treble clef. Pianists achieve virtuosity by complete mastery of a very difficult execution.

Jazz pianists have a different approach. They play from a "lead sheet" that uses only a single melody line and one or two chord changes per measure to describe a song. The lead sheet is the executive summary of the song: It contains the minimum amount of information to express the musical idea. In this way, the song provides flexibility to the artist. It's never played the same way twice. Jazz musicians make their own "arrangement" (specific notes, timing, and tempo) and improvise around the basic framework. The famous jazz pianist Thelonius Monk achieved his unique rhythms by the way he and his drummer spaced their notes around and off the basic structure of the song.

Automation, pre-AI, takes the classical pianist approach. Every action or rule is explicitly programed into the app. But AI developers take the lead-sheet style. They outline important variables or data elements in a model or algorithm. When these update automatically, the algorithm can take on different shapes and patterns. The prediction or decision isn't predetermined but is rather based on calculated probabilities.

This seemingly sophisticated approach leads to less predictability. The game show *Jeopardy* that aired on February 16, 2011, had an unusual contestant — IBM Watson. Watson easily defeated its human competitors, including Ken Jennings, the most successful champion in history. But along the way, Watson made a surprising mistake that left the audience gasping.[9] In the "Final Jeopardy" category "US cities," Watson incorrectly answered Toronto, a Canadian city.

Why would it guess that Toronto was a US city? IBM engineers had to explain the error.[10] While searching through data, Watson found two facts that connected Toronto with the US: Toronto is a North American city, and its baseball team, the Blue Jays, play in the American League. Thus, Watson thought that Toronto was in the US. In answering the question, Watson's confidence level was only around 30%, but it was the best answer the AI could offer.

Obviously, that example wasn't a matter of life or death. However, that wasn't the case for Indonesian Lion Air Flight 610, which plunged into the Java Sea in 2018, killing all 189 passengers and crew.[11] The "angle of attack" sensor insisted the nose of the plane was too high. Automation took control, sending the plane plummeting. The pilots fought the machine, and the machine won. The 2019 air crash in Ethiopia is eerily similar.[12] Even before the Boeing 737 Max disasters, it was evident that we don't understand the risks introduced by increasingly automated vehicles.[13] In 2013, the Federal Aviation Administration found "unexpected or unexplained" behavior of automated systems present in 46% of its accident reports.

Here's the point: Algorithms that drive cars or make loan decisions will increasingly have "Toronto moments." Humans will work side by side with robots that make decisions, sometimes with incomplete data sets or random signals. To put it bluntly, the machines won't always be right.

The Black Box Expands

AI will lead to unpredictable events and challenge us explain them. Forrester has spoken with enterprises worried that their business will turn into a black box, viewed solely in terms of its inputs and outputs but without any knowledge of its internal workings. They fear that the black box will fill with self-enhancing algorithms that defy their understanding. In fact, we now use words like "opaque" and "transparent" to describe a new automation.

These words have a simple but ominous meaning in AI. Transparency is good — we know how the thing works. Opaqueness is bad — we have no clue. Deep learning based on neural networks, for example, is proving more accurate than previous algorithms but is the most opaque. Layers of computation toss away data to get to desired outcome. The door opens wider for unforeseen events, bias, and misuse.

Training bias is one of AIs biggest challenges. Good decisions depend on the data used to train the system. There are well-publicized cases of police-operated computer programs that discriminate against African Americans or identify any person with a high rate of blinking as Asian. A machine might rightly conclude that most programmers are male. Sad, but currently true. But it might also conclude that males are better programmers when, in fact, the training data reflects a forty-year-old view of education.

We'll need constant checks against biased algorithms. Because this is becoming so important, an enterprise software market is emerging to provide "explainability" for AI systems. For example, in May 2018, Pymetrics open sourced its tool for detecting bias in algorithms.[14] We'll need it.

The Tech Industry Won't Protect Us

Many have warned that AI could lead to unforeseen consequences. Professor Stephen Hawking warned that AI could mean the end of humans. He said: "It would take off on its own, and re-design itself at an ever-increasing rate. Humans, who are limited by slow biological evolution, couldn't compete, and would be superseded."[15] This unpredictability stems from a machine's potential to reprogram itself and grow smarter on its own. This is decades away, but a more pressing fear drives from the attitude surrounding AI, best characterized by Mark Zuckerberg's comment, "Move fast and break things. Unless you are breaking stuff, you are not moving fast enough."[16] Eric Schmidt is the former CEO and chairman of Google and now a technical adviser to Alphabet. He has a similar view:

> The world is so competitive and moving so fast. The tech company perspective on releasing new products has to change. For most of my career at companies like Intel, software was released in a very structured way. A disciplined approach grew out of the major mistakes that were made. You had long periods to figure out what the requirements were, then you had time for development, testing, and different stages of release. No more. Tech

had moved to rapid release approaches. Basically, if you think you need time to finish your product then you will lose.

To conclude: We've gotten comfortable with a tech industry that moves fast and plays loose with its products. The term "vaporware" refers to the common practice of announcing and deploying software prematurely, knowing it will be fixed later. That's not a good AI practice. That first release of software is like the first pancake off the griddle. Maybe the stove temperature wasn't quite right or the frying pan's surface is a bit sticky. The chef might be unsure of when to make the flip. The second batch of pancakes is always better, and it's the same with software. But this loose approach to software isn't acceptable for AI. Applications in healthcare, the military, and finance, for example, won't tolerate the risk for AI systems. Collectively, we need to think through the potential consequences.

Data Insight May Cause Blindness

The story of David and Goliath tells of a smaller and weaker underdog who faces a much bigger and stronger adversary. The Goliaths today would be Facebook, Amazon, Netflix, and Google, the so-called FANG companies. They're getting more powerful every day. Google accounts for 70% of online searches in the US.[17] Its search engines improve with each query, some 30-plus billion per month across platforms. Facebook has more than 2.3 billion active users.[18]

The rest of us are David, and as Wilt Chamberlain, the first dominant 7-foot basketball player, said years ago, "Everybody pulls for David, nobody roots for Goliath."[19] Today, few of us are rooting for the FANG companies, yet we're happy to keep feeding them. We offer up our most sensitive personal data in exchange for a little bit of convenience. Connections to friends, ride-sharing ease, or the fun of online shopping is all it takes. We pour in baby photos, vacation pictures, our locations, and political opinions. At first, we had a fairer exchange. The smartphones in our pockets became awesome tools. We loved the instant access to product reviews, comparison sites, and unprecedented information. We could bypass traditional companies and make deals with the sharing economy.

All that's good, but over time, the value equation has shifted in Goliath's favor. AI algorithms have made the data we give up more valuable. Click on a "like" or provide a reference and, bingo, you've created value. Digital profiles now emerge from our browsing histories, social media posts, and location tracking. And we're just at the beginning: Cameras and microphones have moved into our kitchens and bedrooms. Twenty-five percent of

American households already have smart speakers like Amazon Alexa or Google Home.[20] Eavesdropping to serve up targeted ads is already a concern.[21] The privacy worry will be with us forever — from here to eternity.

And it's not just the Goliaths. Hundreds of technology firms can see, use, and profit from data that a customer uploads to their platforms. Imagine you're a shipping company and you buy electric trucks from a supplier. It offers you interesting maintenance and efficiency services based on your fleet behavior. So far, so good. But what if your biggest competitor also buys trucks from the same company? Will it sell your competitor some predictive service based on data pulled from your activity? You bet it will.

Here's what this means: Control will shift rapidly to enterprises that deploy AI solutions with access to relevant data. Many customers and citizens will begin to question the value exchange with Goliath. They'll fight to get back a portion of value for the data they share. And this goes double for governments. China already uses AI to monitor and control citizens, and it has a goal to build a $150 billion AI industry by 2030.[22] During his 2018 New Year speech, the president of China, Xi Jinping, had the 2017 book "The Master Algorithm" visible on his bookshelf.[23] The book hints that an uber-algorithm could be developed to link the most important machine learning methods together, a decent James Bond plot. The point is simple but important: AI gives heightened power to those that understand and deploy it. We're just beginning to explore needed checks and balances.

Most of you probably know this story already. Facebook's Cambridge Analytica data debacle is the best or worst example, so far, of data misuse. Here's the short version: The firm developed an app to survey a user's personality traits.[24] No problem there, but it then combined the results with "friend connections" for 87 million people. The machine could then predict your voting tendencies. Conservative strategist Steve Bannon used the results to push voters toward his populist vision for the US and to help elect Donald Trump. It wasn't until major news outlets exposed Cambridge Analytica's actions in 2015 and 2018 that Facebook publicly reacted.

Governments are enacting regulations like the European Union's General Data Protection Regulation (GDPR) worldwide to try to protect our data privacy. But can governments be the countervailing power against a hard-charging tech industry they barely understand? If the April 2018 Facebook hearings in the US are any indication, the answer is no.[25] For two days, Mark Zuckerberg testified in front of Congress in an attempt to soften the Cambridge Analytics scandal. Some elder senators were alarmingly clueless about social media and the tech industry. But Senator Richard J. Durbin of Illinois got to the heart of the issue. Politely, he asked if Mr. Zuckerberg would tell them what hotel he stayed at last night. Zuckerberg didn't see where this was going, stumbled a bit, and then stammered, "No." The senator continued, "If you've messaged anybody this week, will you share with us the names of the people you've messaged." Zuckerberg finally got it.

Some 45 times, over two days, the Facebook CEO assured us we have control of our data.[26] He referred to the 3,300-word user agreement, written by lawyers for other lawyers to read. He noted the dozens of places where you can make public versus private decisions on what others can see, if you took the time to figure it out. But here's what he didn't say: The good stuff is behind the scenes, anyway — the ads you clicked on, the communities you take part in, loyalty card usage, and biometric data from your photos. There are no user settings for these.

The scale of these great data collection pools is staggering. Social media platforms are exploiting the human weakness for continual social or business validation and using it to collect data at unprecedented scale with limited effort and cost.[27] Monetizing the natural and effortless data collection is a certainty.

Workers Lose Control Of Wages

For three-plus decades after World War II, the US built the largest middle class in the world. Hard work paid off and resulted in higher wages and better lives. Hope and optimism prevailed. But the past 30 years have reversed this pattern. Even with the economy doubling, wages for the middle class have grown only modestly. Confidence that the economy will provide high-paying jobs and improved living standards for the majority has declined.

The big question will be, "How will people make a living wage?" Indeed, the cost of living in many places in the US has long been rising, while incomes remain stagnant. The extent of the problem was laid bare in 2003 when the MIT Living Wage Lab created a tool to estimate the minimum living wage across many US cities, including housing, groceries, and basic healthcare.[28] Calculations from the tool show that in 2017, the median living wage in the US was $16.07. Today, 42% of Americans make less than $15 an hour, and 30% make less than $10 an hour.[29]

There are two main points here: Firstly, future-of-work studies focus too much on job loss and not enough on workers' ability to control wages and working conditions. And secondly, automation rewards are going to a small percentage of investors, executives, and digital elites at the expense of nearly everyone else. AI can solve hard problems and make jobs better for many, but companies must invest in these opportunities. If history is any indication, the primary use of automation will be to save money and return profits to shareholders.[30]

Automation Shifts Control To The Digital Elites

Digital elites are grouped as one of our 12 work personas. They are educated, workaholic, entitled, and connected. They tie their success to continued automation, and they'll be the big winners. AI conferences are filled with them. Bill James (not his real name) is a regular guy, not a tech founder or senior executive but one of thousands with a promising future in the tech industry. His firm develops software robots that replace humans in back-office jobs like accounting. He whips his phone out like a cowboy and slides it across the table, putting his fellow guests on probation. If they don't keep his attention, he has plenty to do. He'll check his phone and ignore those losers. Bill explains:

> I'm a sales and marketing guy. I tend to exaggerate and don't go very deep on the tech stuff, but here's what I do well. I match a basic understanding of a technical capability with what a customer needs. I really have no idea what a neural network is, or deep learning, computer vision, but you sprinkle some of these terms throughout a pitch, it really works.

> I've been in tech for 20 years and I've never seen a lower bar for truth than the conversation around AI. The marketing buzz is just insane and generates a lot of confusion about what it is. My PowerPoint slides paint a cool picture, and if a prospect challenges me about an AI connection I just say, "Well, yeah that's true but this is only a PowerPoint slide." Maybe in today's world, words don't matter as much as they used to. Somehow, dumb software products that have been around for years are suddenly brilliant.

> Oh yeah, the jobs thing. I don't dwell on it too much but a lot of the software we install does stuff that people hate to do, like creating reports or updating some dumb database. They should go do better things, and if we don't automate their work someone else will anyway.

Here's the point: thousands of "Bills" in the tech industry take a utopian view of the future of work. Automation is all good. Humans are freed from being robots. They soft-pedal the effects of job losses or stagnant wages or just don't think about them. No company has done more to push AI than IBM. This comment from its CEO, Ginni Rometty, at the 2017 annual World Economic Forum in Davos, Switzerland, is typical of the business viewpoint:[31]

> It [AI] is a partnership between man and machine, if you
> want to put it that way. Think more about activities
> changing with the technologies. When you do your job,
> there will be things that take you a lot of time to research
> do. Yes, they'll be done faster. Then you have the time
> to do what I think we [. . .] humans do best.

Practitioners like Bill James, as well as many top industry executives, think all this is fine. AI will be a benign, liberating force that leads to more rewarding work. Jobs will be better. But as we'll see in later chapters, there's a numbers problem. Workers in select personas will lack the digital and human skills to transition to better jobs. Some will become cloud engineers, robot nannies, or even data scientists. But millions of workers will flounder in the gig economy or become automation deficits — unless we do something about it.

3 A NEW FORMULA FOR SCALE

Scale drives profit and revenue and has always been a corporate objective. College-level microeconomics textbooks praise economies of scale to drive down marginal costs and lower the cost of the next item produced. For hundreds of years, scale required three things: more land; more capital; and, in most cases, lots more people. In 1955, Boeing used nearly all its profit since the end of World War II to produce its first commercial jet.[32] In August of that year, fuming CEO Bill Allen watched his Boeing 707 prototype do an unplanned barrel roll. GM scaled with a mix of all three. In May 1979, its US employment peaked at 618,365, making it the largest private employer in the country, with worldwide employment of 853,000.[33] But that was the old version of scale.

Digital Scale Requires Fewer Workers

Digital scale grew from our publicly funded technology infrastructure. Consider, for a minute, the fruit of your tax dollars. The US Defense Advanced Research Projects Agency (DARPA) funded the internet.[34] Google developed its algorithm with funding from the National Science Foundation.[35] The US Department of Defense funded GPS and Siri and the nonprofit R&D centers that support its mission.[36] The tools of digital scale — cloud platforms, mobile, social media, and AI — emerged from these public investments.

Digital scale has a different balance of land, capital, and labor. One ingredient has become less important, however: humans. Enterprises still need digital elites with skills in social media, data science, machine learning, and specialized automations, but the new form of scale doesn't have the balance of labor that previous automations did. Look at today's most valuable companies. Facebook employs only 25,000 people and generates a staggering $1.6 million in revenue for each of them.[37] Apple generates $1.85 million per employee, and Google parent Alphabet generates $1.2 million. Traditional scale added people to manufacture, sell, and fill orders. Now, even new brick-and-mortar stores can open with no people. The Amazon Go store, for example, uses computer vision, facial recognition, sensors, and mobile apps to eliminate humans from the store. It uses no cashiers; no one stocks the shelves; and there are no scary guards to avoid. The store is managed via a dashboard from an office park. Enter the store through a gate that opens with the Go application on your mobile device, pick up what you want, and leave — that's it.

Even capital is less important. Who needs to open up a factory to create a business? Many native-born digital startups begin by simply opening a laptop. A digital business can scale with low- or no-cost customer interaction and digital pathways. Airbnb and Uber own no property or cars, lease their office space, and consume needed technology as cloud services. Capital; land; and, particularly, people, aren't fundamental to expanding their businesses.

The FOA Drive Operational Scale To New Heights

The FANG-type businesses we describe above were digitally born, and there will be many more of them. But can traditional businesses scale in the same way? They can. Online commerce certainly has, and in turn, has created a modern form of ruins: abandoned shopping malls. These ruins don't inspire romantic sentiment, but you can visit a web site dedicated to them.[38] And where did all the people go who worked in them?

Automation has fueled the expansion of chain stores of all kinds and will continue. Drive down almost any highway, and scaled businesses dominate our 21st-century landscape. Retail food franchises like Dunkin' Donuts, Popeyes, and Taco Bell, as well as box stores, pop up like weeds on a new lawn. Bank branches, chain stores, and retail coffee outlets spread without resistance through once-vibrant neighborhoods. They bring a placid sameness, standardized decor, and a lack of any individual vision or creativity. To veteran urban dwellers, this form of scale strips vitality from their cities. Local specialties vanish. Small, independent shops; galleries; bookstores; antique shops; and lifestyle businesses of all kinds can't make the rent. The environment becomes convenient, low-cost, and dull.

Those are the lifestyle aspects of scale, but what about the workers? Let's start with McDonald's, always a leader in operational scale. When it comes to fast food, it wrote the book. More than 34,000 restaurants serve 69 million people in 118 countries each day.[39] It sells 75 burgers every second. All this activity generates vast amounts of data. In the past, McDonald's produced summary metrics on a per- restaurant basis, which made it difficult to pinpoint problems in specific areas such as drive-through or ordering patterns. But now, a data-driven metrics culture has taken over. In-store traffic, data from point-of-sale (POS) machines, video, and sensors is analyzed with machine learning algorithms to spot trends. The restaurants all look the same, but data now makes each unique.

A director at one of McDonald's oldest franchises can tell us how automation is contributing to scale and the effect on workers. Ray Kroc himself recruited the original franchise founders, who now operate multiple restaurants in the New York metropolitan area.

Yep, we have become data driven. There is nothing we do that you can't find a metric for. They are black and white but ignore grey areas. And you lose the full story with the metrics. Maybe that store had a string of bad customers.

Headquarters now calculates the amount of sales per employee, not at the end of the month or week, but in real time. They tell the franchise partner to reduce staff when sales drop below a certain level. A worker might be traveling in, maybe on a bus, and sometimes we tell them not check in for an hour or so until more customers are predicted. We try to give them enough hours, so we don't lose them. It's a balance.

But here's the thing. The metrics make the workers nervous. If someone underperforms, they will be very unhappy. And sometimes they shuffle the execs at corporate and a new guy will change the metrics. Store managers and employees scurry around and try to adjust. It is not a relaxing place.

Food use to be the biggest cost but now it's labor. Automation investments will target people. Corporate is now working on robots to make French fries, but we won't see Flippy the robot anytime soon. In fact, we don't flip burgers at all. A 300-degree plate lowers to the stove to cook the other side of the burger. Sometimes automation improves the employee and customer experience. We have kiosks in 25% of our restaurants. Employees now help customers and have time for table service. We worry about Chick-fil-A's reputation for a customer friendly environment.

Banks also are also under tremendous pressure to scale. This can lead to rogue behavior and put more pressure on workers. To start with, the large banks just aren't warm and fuzzy places to work. Let's look at bank tellers. Some 80% of 500,000 are women, and three out of four earn less than $15 an hour. Here's Cindy, a teller with a major US bank:

I have to buy my own [clothes] and look professional every day. New scheduling software tells me when to come in next. I used to have regular hours, but now as a single mom I have trouble managing my day care and other errands. But what gets me is the way I'm watched. At the window, I must make two product offers to every encounter. The person's name must be mentioned. I feel like I'm constantly being judged.

I originally took the position because the job description made it seem like a lot of fun. There was no mention that I would be required to cold call customers and force products and services down their throats. The "aggressive goals" were constantly being changed and our requirements became more and more. Higher ups do not care about the individual employee nor do they care much for the customer. It's all about numbers and "growing the branch."

Here's what we can take from this glimpse into banks and retail food: The pressures of scale will drive companies to develop data-driven cultures and, if unchecked, will be of growing concern to workers. We're already seeing the battlegrounds forming. The State of New York's Department of Labor now forces companies to pay a fee to workers for what's commonly identified as "just in time," "call in" or "on call" scheduling.[40]

Here's a second point. We need to pay these tellers more. Bank tellers like Cindy now handle more difficult issues, like why a person-to-person electronic transfer didn't work. The easy transactions have been automated, but robot tellers won't replace the human ones anytime soon. HSBC has a finance robot called Pepper.[41] It's sparkling white and about 4 feet tall, with large blue eyes that light up and a tablet attached to its chest. But Pepper can handle only the most basic questions. Pepper's primary skill is to lure street traffic into HSBC's flagship branch on New York's Fifth Avenue.

AI Drives Scale From The Bottom Up

AI investments will start to eliminate middle managers, cubicle workers, and even university professors. Blockchain, RPA, and online education platforms are just a few examples of where AI is driving scale.

Let's start with middle management positions. Adding more humans to manage operations just hurts a companies' ability to scale. It's much better to let algorithms be the boss. And it's already happening: A million Uber drivers in the US and Canada have no humans or coworkers to speak with when something goes wrong.[42] They're completely controlled by an algorithm that watches everything — ride acceptance rates, trips completed, hours spent logged into the app, ride cancellations, and reports on driving behavior. Predictive analytics relocates drivers to take advantage of "surge pricing." For Uber drivers, the workplace is a world of constant surveillance.

In the next few years, algorithms will replace many managers who perform coordination functions. Today, some 36 occupations and more than 10 million workers fit into the coordinator persona. They manage human tasks and schedules, orders, and shipment logistics, but algorithms can easily replace them. If you have a clipboard in your hand, you'll soon be putting it down.

Here's why: Software bots and machine learning algorithms will run on servers to schedule that furniture shipment, communicate with field workers, or post data to an order system. The math is simple and frightening. Already, a software bot, built with an RPA platform, can cost $15,000 to maintain but can replace two to three humans doing coordination functions. Do we think enterprises are ignoring these cost efficiencies? They're not. The top three RPA platform vendors alone have 5,000 business customers that have already invested in their software platforms.[43]

Coordinators are also threatened by automation that speeds

communication. Collaboration technologies like Microsoft Teams or Slack promote peer-to-peer communication and reduce the need for a human to keep everyone on the same page. The result? Flatter organizations and fewer middle managers.

Software Robotics Inch Into Cubicles

In the 1960s, a secretary might type quietly outside your door or enter your office to take dictation, just as in the television series *Mad Men*. In time, these secretaries moved to a typing pool. By the 1990s, word processing had reduced their numbers. And by 2010, personal productivity apps like calendaring and collaboration had largely finished them off. Automation has consistently altered administrative and office support and will continue to do so. Chatbots that become virtual agents and software robots that do background tasks will slowly eliminate today's contact center workers.

There's no question that software robots will start to empty out cubicles fast. Comparing the 12 personas, cubicle dwellers are, by far, the workers who will be most devastated by invisible robots. This is no small matter. Today, more than 20.7 million people in 72 occupations work in these positions. Holly Uhl, a manager at State Auto Insurance Companies in Columbus, Ohio, oversees an automation project for back-office tasks that has eliminated an estimated 25,000 hours of human work a year.[44]

Blockchain Targets The Coordinators And Cubicle Workers As Well

Blockchain was derived from open source technology that spawned Bitcoin, the first cryptocurrency. Blockchain stores data as "blocks" linked together to form a "chain"; the more transactions recorded, the longer the chain. The assets can be tangible, such as agricultural products, or intangible, like Bitcoin or Ethereum currencies. Bitcoin and other cryptocurrencies are just one of many applications that use the blockchain framework. With a blockchain, transmitting information, transferring funds, or approving a transaction can all happen digitally.

Blockchain has no central authority or administration; the architecture is decentralized and nonproprietary. Sound familiar? It should. These characteristics are the same as another nascent infrastructure first developed in the 1960s: the internet. And, like the internet, it has the potential to expand virally.

Here's why: Ship a container of avocados from Kenya to Italy, and you'll involve some thirty entities. Port clerks, shippers, buyers, and government officials shuffle paperwork and enter data into legacy applications. Paperwork, and the humans to manage it, represent a minimum of 20% of the cost of the avocados.[45]

Automated solutions based on blockchain will store "smart contracts" in the open source digital ledger.[46] Participants will have instant and secure information, driving out paperwork and signatures and changing the roles of the humans that shuffle them. But there are factors that will slow blockchain's progress. Here's Martha Bennett, VP and principal analyst at Forrester, an international expert on blockchain applications:

> To start, smart contracts are neither smart nor contracts in the legal sense. "Smart" implies use of AI or advanced analytics; the truth is, blockchain provides none but is really only a trusted form of digitization. The same could be accomplished for many of the hyped blockchain use cases with a database, some workflow software, and email. In terms of jobs, blockchain, in theory, can replace many clumsy processes, but humans need to agree to change. Standards for data formats, use, governance, access rights, and IP management all need to be decided. And this takes time. Labor dynamics will also slow blockchain. In the shipping logistics example above, the middlemen — the authorities involved — are a protected workforce.
>
> That said, many forward-looking shipping firms and authorities plan to use blockchain to streamline processes that span multiple organizations. Here's another perspective on blockchain in the context of automation: what ERP and BPM have delivered for intracompany processes, blockchains and smart contracts can deliver for multiparty processes (i.e., spanning corporate boundaries).

Martha tells us this: Contrary to what you may hear, blockchain, in the near term, won't replace intermediaries or existing trust networks but will rather change their roles. Blockchain won't affect our job numbers for five years or more.

Ivory Towers Grow In The Cloud

Scale has become an obsession for all enterprises, even for nonprofits. Leading universities, for example, are catapulting toward an impersonal, but profitable, form of education. Cloud support, mobility, and social media intersect to provide a new way to scale. Schools now use these resources to acquire and service more students at lower costs, meaning fewer and lower-cost adjunct teachers.

For top-tier universities, it's a difficult line to walk. Online education doesn't have a sterling reputation. Commercial ventures like Trump University and the University of Phoenix have been compared to boiler-room operations that use predatory recruitment practices to encourage students to pile up government-backed debt for near-worthless degrees.[47]

But new approaches are gaining momentum. The 2U online program manager (OPM) approach, for example, partners with top universities. This gives credibility to the online degree. The university will "white label" 2U's cloud-based software and split the tuition payments with it. The best part? 2U leaves all academic content and delivery decisions to the real university.

Online education draws passionate debate. The attitudes of hiring managers, the quality of the education, and the price points top the list. But the effect on the future of work leaves little to question. There will be fewer tenured faculty and lots more part-time adjunct positions. The adjuncts come from the talent economy, often from top jobs, and many have great degrees, but they don't do research. For most of them, teaching is a sideline, not a lifetime ambition. But the economics favor them. The University of North Carolina doubled its business school enrollment to more than 900 (twice as many as were in the traditional MBA program) without adding any buildings or faculty.[48] Obviously, this form of scale in education will reduce the need for full-time teachers.

But is this progress? Al Arsenault has a view and is qualified to share it. He was a recent student in an online master's degree program but also earned a "traditional" degree in residence at Purdue University. He taught college classes at the US Air Force Academy and, for 11 years, was an adjunct professor at the University of Maryland, Baltimore County. Al tells us:

> Scale is what's going on here. Berkley had a sleepy Department of Library Sciences. A few years ago, they renamed it the "School of Information." This helped a little, but to grow they needed something else, and they

had no budget for whatever that might be. They partnered with 2U and grew quickly from 30 students to 300, all online and with adjunct instructors.

You can divide online education into synchronous and asynchronous experiences. Synchronous gives you online office hours and at least some online human interaction. At Berkley, which was a 2U system, we used Adobe Connect and now Zoom for live audio and feedback. We even had a Slack channel. They at least tried to give us some of that on-campus feeling. There was real interaction with teachers and students. We attended the same graduation as the on-campus degree students. You could not walk down the hall for the casual talk, or have a beer with fellow students, but it was pretty good.

But asynchronous is a different story. There is no interaction with teachers or other students. My daughter had an "asynchronous" experience for a master's degree at GW. It was a lot more expensive than we thought. She had no shared projects with other students. No one-on-one chat time with teachers. There were videos to watch, papers and books to read, assignments to do, but she never met anybody in person. She didn't attend graduation. What would have been the point? She wouldn't have known anybody. This lack of human connection won't hurt her in the short term. She will have her degree. But what she won't have is a network of colleagues and mentors. People her age need to build a personal brand. That's where work is going. And online education doesn't move that dial.

Here's the takeaway: It's not just getting the degree that matters, or even what you learned. Traditional education helps build a network of people. When you speak with very successful people, they talk about those that helped them along the way. The ones that sent the elevator back down to let a younger person on — an inspired teacher, a mentor in an early job, the roommate in college, or an extended community of teachers and students. Asynchronous online education hinders this network development. But more synchronous approaches don't have to. The answer may be to add a few more teachers, take a bit less profit, and adapt to a synchronous model, one that uses the latest collaboration technology.

4 CONVERGENCE BRINGS THE PHYSICAL AND DIGITAL WORLDS TOGETHER

The physical world is all around us. Products, components, and equipment in cities, worksites, and factories have been around for a long time. In comparison, the digital world is new, made up of digital health records, bank accounts, social media, and sensor-generated information. When these two worlds come together, convergence results. Important advances for convergence are IoT solutions; an ever-growing network of physical objects that have an IP address for internet connectivity; and, more recently, edge computing, which brings memory and computing power closer to the location where it's needed. These are converging the physical and digital worlds at a blistering pace.

The goal for most convergence efforts is to enhance physical experience with digital insights. Visit a Disney park and get your MagicBand, a wearable device to navigate the theme park and speed access to rides, restaurants, and hotel rooms. IoT-enabled building and facilities management targets commercial offices, warehouses, and healthcare facilities. Efficiency of energy, space, or operations is the primary goal.

Self-service kiosks, intelligent POS systems, digital signage, and intelligent video analysis are examples of a converged physical and digital world. Customers will have personalized interactions in retail stores, hotels, public parks, and industrial sites. For example, retailers can identify a physical area where a consumer is willing to buy and personalize the experience. The goal? To use spatial smarts to make for an exciting experience, one that Amazon couldn't easily replicate.

Our cities will be big winners. They'll get smarter. Garbage cans will signal when they need emptying. AI-based monitoring and surveillance will reduce crime. Sensors in public transportation will allow citizens to track wait times and managers to improve routing. City-provided bike sharing is exploding. Cities are trending to be cleaner and less scary than ever.

This is all good, but where does it leave our workers? A few rapidly advancing areas can give us a clue: In Japan, robots are replacing elder care workers; security workers will lose millions of positions; building facilities staff will be fewer in number; we'll see fewer logistics managers caring for shipments; predictive analytics and proactive maintenance will reduce equipment repair teams; and automated vision systems will replace human inspectors. Let's drill into a few of these.

Security And Surveillance Doesn't Have A Rosy Job Outlook

Security guards patrol and protect property against theft, vandalism, and other illegal acts. This job category employs more than 1 million US workers who earn between $16,000 and $24,000 a year.[49] Armed guards, bank guards, bodyguards, bouncers, security officers, and transportation security screeners fall into this category. The US Bureau of Labor Statistics predicts "excellent overall job opportunities" for security guards through the year 2026, with employment growing at about a 6% rate.[50]

Forrester's numbers are very different. Protective services are one of 70 occupations in the location-based worker persona, which includes more than 29.9 million jobs. We do forecast some job gains. For example, technicians will need to manage emerging IoT surveillance systems and repair security robots, but overall, we see far more automation deficits. The numbers for security and surveillance, like some other BLS forecasts, seem unrealistically optimistic once we factor in the progression of automation.

Look around you. There are cameras everywhere, and they're getting smarter. Physical security will use devices such as card readers, door locks, mobile robots, and cameras as well as card personalization software and analytics. We'll see reduced numbers of security guards in government, retail, healthcare, manufacturing firms, and office buildings. These jobs offer employment to many people with limited education and experience. They might not do much for one's self-esteem, although a clean record is a primary requirement. Olek Sumoski, for example, came to this country from central Europe.[51] He finished high school and has had various security office positions. It's fair to say he's developed a bit of an attitude:

> It's a thankless job, sadly. You get mocked for it, yeah.
>
> Failed cop, etc. It doesn't affect me, since I never wanted
>
> to be a cop, anyway.

> If you work at a store, customers might accuse you of
>
> following them, when really, you're just walking around
>
> because you're bored. Store employees expect you to
>
> follow thieves around when you're just there to observe

and report. It's annoying to hear that you're not being "useful enough."

If you work at a bank or shipping facility, you'll have to check ID badges or personal bags. Employees will give you a hard time by not showing them to you and walking by, thinking they're being cute and defiant. Believe me, a security guard can get you in more trouble than you can us. We have periods of time where we can get even with you. We signed up for the job knowing what we are instructed to do. We hate it, you hate it. Deal with it.

I'm sure they're making robots who can do it my job. But I worry about the armed positions; humans decide today if they should put a bad guy in his place, a decision with lots of potential grey areas. Robots are ones and zeros, black and white. How will they decide?

I liked doing sites without having to face the public. It's more boring, but damn, just treat us with respect.

Olek tells us two things: Firstly, these aren't great jobs. People, if they can, will try to do something else. Something that's more interesting and that gains them more respect. And, secondly, automation will alter these positions over time, but employers shouldn't undervalue the human judgement they require.

Granny, Your Robot Is Ready

The growing elderly population will expand jobs for human-touch workers, one of our 12 work personas. But will growth for this persona keep up with worldwide demand? Apparently not in Japan, where a shrinking labor force and exploding demand for elder care is at crisis level. By 2025, the country will be short 380,000 elder care workers, though others put the figure closer to 500,000.[52]

But Japan has a plan to deal with the elder care problem. It's massively funding robot development, a commercially exciting area. This type of coordinated approach gave Japan leadership in the machine tool industry in the 1980s. The US, by contrast, has no coordinated technology funding but instead scatters tax dollars to defense contractors and universities. Occasionally, successes like the internet emerge, and that was a good one. But the lack of a coordinated policy may put the US at a disadvantage for beating China and Japan in the automation battle.

Elder care robots aren't in the distant future. About 500 Japanese elder care homes already use SoftBank Robotics' Pepper for games, exercise routines, and scripted dialogues.[53] The government funded Panasonic to develop Resyone, a bed that splits in two, with one half transforming into a wheelchair. Cyberdyne's HAL — short for Hybrid Assistive Limb — is a powered exoskeleton suit that helps caregivers lift people. Those needing walking rehabilitation can grab hold of Tree, made by Reif, which crawls along the ground and offers balance support. It's fair to say that Japan is all over convergence opportunities to solve domestic problems, which, in turn, will create an export market.

The care homes must resolve some significant obstacles: high costs, robots that are too rough for fragile-skinned elders, and potential harm due to reduction in human contact. But here's the obvious question — wouldn't it just be easier to import care workers from less-developed neighboring countries with high unemployment and struggling populations, like Indonesia and the Philippines? Many would say yes, but it's not Japan's way. Japan wants technology, not people, to solve its elder care problem. To restrict immigration, it created its licensing exams in Japanese only, although signs point to a potential loosening of this type of immigration policy.[54]

Convergence Will Expose A Glut In Commercial Office Space

In the US, we've overbuilt retail space, and now we're in trouble. Online shopping is emptying out retail locations. Dead malls are now a thing. But less highlighted is the potential glut in commercial office space. Today, companies have too much space. The reason: The future of work is mobile and involves borrowing, instead of owning, talent. Companies will need more flexible options than a long-term lease can provide. Emerging IoT solutions such as spatial intelligence will help. For example, in 2018, Microsoft introduced Azure IoT spatial intelligence capabilities that allow better management of heating, cooling, and room-booking systems, based on how customers are actually using the space.[55] In this case, digital smarts are embedded in walls, ceilings, and equipment. Here's a former architect who's now a smart building expert for a leading technology company; he describes the company's recently launched IoT solutions in the area of spatial intelligence as well as its potential affects:

Here's the big shift. The primary sensor we have is in our hand or pocket, our smart phone. Basically, one physical device connects you to the digital world. Fast forward a year or two and the entire space around you will be transmitting data. Instead of a single device there will be a full "space" connectivity. The various sensors connecting the HVAC, lighting, will access your device via Bluetooth and transform the space based on your preferences.

Lights will dim, temperatures will lower. I just left a conference room designed for only dozen people that packed in twice as many. It was red hot. Why not have sensors and analytics adjust the temperature based on room occupancy?

The bigger trend of this convergence of the physical and digital worlds is better space utilization, understanding what actual usage is, energy management, and a push towards office as a service. We need to start by connecting the systems in the building together, to get a holistic view.

Here's the point: Office buildings, manufacturing plants, and retail sites will become smarter. Intel, for example, installed 9,000 sensors throughout its 10-story, 630,000-square-foot office in Bengaluru, India. These sensors collect temperature, lighting, humidity, electrical usage, and occupancy data.[56] Healthier and safer work environments and better visitor experience are the result, but the buildings require fewer humans. Facilities staff will be decrease as well. Those remaining will need updated skills to use a "digital twin," a virtual model of the environment. The workers who support commercial office space, brokers, construction, and service will be reduced.

Convergence Will Reshape The Utility Industry

Utilities in the US employ 555,000 people who span almost all work personas.[57] Smart grid technology, such as that being developed by Pacific Gas and Electric (PG&E), will analyze power usage, monitor the grid infrastructure, and proactively identify outages.[58] This will reduce location-based workers like meter readers, the coordinators who manage them, and the physical workers making repairs. Chatbots that receive orders for a change in service and software robots that do bookkeeping, accounting, posting, and auditing will reduce cubicle workers.

These automations are deploying today, but more dramatic forces are circling. They're called microgrids, and within 10 years, they'll allow individuals to generate their own power and even sell it to others. Out of the 122 startups in the energy-blockchain space, 22 focus on peer-to-peer energy trading.[59] Are utilities preparing for this future? Delores is a VP in the finance department of a New England utility provider. She says:

> I came from the finance industry and have only been here for three years. I kept talking to them about getting disrupted. You know . . . by clean-energy technologies, and people generating their own power and leaving the grid. They would look at me as if I should have a white coat on with constraints and be hospitalized [laugh]. Like I was crazy.

> They would say, "We have been doing it this way for 100 years and we will be here 100 years from now." I'd think to myself, they are making the case for me. If there was ever an organization just waiting to be disrupted, this was it.

> We are so laden with legacy systems, burdensome regulation, and the need to maintain dividends. No way these guys will build the next-generation smart grid technology. Just like the candle makers were unlikely to invent the light bulb. They are more likely to fight clean-

energy momentum with lobbyists and self-serving regulation. Germany, now there's a smart country. They avoid conflicts, separated the old utilities from clean-energy ones and let them fight it out with a level but low-carbon playing field.

Here's the point: Clean-energy jobs to build, install, and maintain wind and solar infrastructure are projected to grow significantly and add automation dividends.[60] Traditional electric power and natural gas utilities, despite being essential to everyday life, will see continual job erosion. Increased size and efficiency of new power plants will help maintain the status quo, but the FOA will be relentless.

Don't Count On Jobs That Manage Logistics Or Supply Chains

Transportation and physical equipment are central to thousands of companies. Fleet managers, asset managers, and various logistics positions are mapped into our coordinator persona. Over time, these jobs will be harder to come by. Take fleet managers. Today, they manage vehicles to deliver and distribute goods, but automation will step in. Remote machines will diagnose engine trouble. Algorithms will make decisions about speed, acceleration, coolant temperature, and brake wear. Here's an example: Navistar, a commercial truck manufacturer, reduced maintenance costs and vehicle downtime by up to 40%, using historical data and machine learning to make predictions.[61] Fleet owners can now monitor truck health and performance from smartphones or tablets, prioritize repairs, and identify the nearest dealer to address issues.

The manufacturing, chemical, railroad, and oil and gas sectors alone have thousands of jobs managing their assets. Convergence of the physical and digital worlds (i.e., IoT systems and edge computing supported by 5G networks) will detect equipment status and schedule work repairs. The result? Fewer humans to manage the process of repairs, fewer repair incidents, and greater potential for remote repairs. The BLS analysis shows growth in these job categories, but our numbers don't. We project a decline in the coordinator persona.

Warehouse, Inventory, Supply Chain, And Retail Jobs Will Struggle As Well

The BLS job forecasts for supply chain management are positive, showing 7% growth.[62] Universities continue to graduate students with degrees in supply chain, transportation, and logistics management. But these studies won't lead to promising careers, for a growing set of IoT and edge solutions will soon manage key components. RPA software robots and chatbots, invisible to the eye, will replace workers in the cubicle worker, coordinator, and function-specific worker personas.

Automation will link partners, customers, and the physical world and reduce humans who provide gateway functions. For example, Denimwall, owner of G-STAR RAW franchise stores in the US, attached RFID tags to inventory items and installed overhead readers onsite.[63] Track-and-trace solutions of all kinds are expanding. ParceLive is a postcard-sized device inserted into a parcel that tracks the location, condition, and security of shipments.[64] Data transmits every hour unless there's an event (e.g., someone drops or opens the parcel, or it exceeds a temperature limit). These delivery-tracking solutions have wide applicability in the retail, wholesale transportation, and logistics sectors.

Strawberries And Digital Agriculture

Farming is one of the world's oldest professions. For some 10,000 years, food was produced primarily with human labor. The 19th and 20th centuries added automation and scale but largely maintained the age-old way of doing things. Automation in the 21st century seeks to converge the physical and digital worlds to reinvent the farm.

Agriculture technology (agtech) startups are suddenly hot.[65] Precision agriculture, sensors in farm equipment, indoor agriculture, and imagery that uses AI to isolate crop issues are segments that savvy venture capitalists are now tracking. Here's why: Sensors are making better use of light and water; lighter, smart machines are replacing massive tractors; and farmers are making decisions based on AI from smart phones. Want an example? Anthony Atlas is head of strategy at Ceres Imaging, a five-year-old precision agriculture company with 60 employees. He explains what they do:[66]

> We have sophisticated cameras that take pictures from an airplane. We use sort of an "Uber like" approach. We don't own planes or drones but use what's available. We upload the data to a machine learning server. Thousands of acres are analyzed to detect problems, patterns that nature would not create, maybe due to a lack of irrigation or an infestation. Climate change is pushing our business as well. Think about it. Your goal as a farmer is to create

38

a consistent yield. How do they do that with the crazy weather we are seeing?

We send alerts and data to the farm. The farmer then goes out and decides what to do. They know every acre of their farm. They accept the value of the technology but are a tough crowd. GPS-driven tractors took years of convincing.[67] It made many farmers nervous. The attitude was like, "I know how to drive the damn tractor." Now almost no farmer drives one.

Here's the point: AI for precision agriculture is now available in the cloud. This allows a small group of scientists and computer experts to do amazing things with food. AI like that at Ceres Imaging can help meet world food demand.

On the human side, we'll soon have a shortage of farm workers to pick crops. This physical worker occupation is one few aspire to. The jobs are dirty, dangerous, and tedious, and the pay is poor. They'd be perfect for robots if robots were ready, but today, a human can pick about an acre of berries a day and put them to shame.[68] People can pluck berries that are hidden behind leaves, drop them into clear plastic packages, and run them to a waiting truck. A robot must use GPS with a high-definition camera to find individual berries. When it gets to a strawberry plant, bright lights flash; cameras spin in a circle; and a robotic claw picks the berries. But the robot has trouble with rough terrain and is clumsy and as big as a bus. And, like many humans, robots don't like rain.

Convergence Helps Diversify The Workforce

Convergence is helping open up jobs, regardless of gender. Traditional male occupations are declining, anyway. Jobs such as locomotive firers will shrink 70% over the next 10 years.[69] Men still dominate physical worker occupations such as agriculture (76%) and construction (93%), but women dominate human-touch jobs such as nursing and teaching, which are both growing.[70]

Female physical workers are also becoming more common. Robots and industrial automation now do the heavy lifting at a plant, for example, or do less desirable underground work that men used to do. Let's peek into an Amazon warehouse. Millions of retail jobs and stores have vanished due to online shopping, but Amazon has become a hiring machine for warehouse workers. Just a few years ago, exhausted workers ran laps to pick and pack orders that often required heavy lifting. Now, Kiva robots (acquired by Amazon in 2016) help. One result? A level playing field regarding gender. For example, at her job, Nissa Scott can now let robots do the heavy lifting.[71] Instead of picking up 25-pound bins during her 10-hour shift, she watches her replacement — a giant, bright-yellow mechanical arm — do the stacking. Convergence will transform many tasks in challenging physical conditions. On construction sites, for example, male workers often lift heavy items from one floor to the next as a one-off activity or carry sheetrock up two flights of stairs.[72] Convergence promotes a more diversified workforce, one that's safer and less physically taxing for all.

5 BLACK MIRROR FORECASTS AND THE PERILS OF THE S-CURVE

Future-of-work studies take polarizing positions. The dystopian view predicts that robots will hollow out the economic core of millions of working-class people and chip away at white collar jobs. This view gets the most press coverage. The utopian view gives hope that all will work out in the end and aligns with many political and commercial agendas. But the future is rarely black or white. Instead, it's a gradual mix of winners and losers; positives and negatives; and unforeseen innovation and human adaptation driven by battling economic forces and, most of all, unpredictable human behavior.

So why are there so many "black mirror" scenarios where the ugliest (blackest) aspects of human nature are magnified, including firms callously discarding workers? Why are there so many predictions of massive job loss and evil robots? In September 2018, for example, Kai-Fu Lee's book *AI Superpowers: China, Silicon Valley, and the New World Order* predicted that within 15 years, AI would have the technical ability to eliminate 40% to 50% of jobs in the US.[73] But, as we'll see in the upcoming chapters, our forecasts factor in automation dividends and take a pragmatic view of AI progression. We predict that only 16% of jobs will be eliminated by 2030

Before we start, forecasts are really hard to get right. They're usually correct about half the time. Forrester has done many, and this is about average. Fortunately, correct forecasts get more attention than those that turn out to be off-target. But a few reasons stand out for missing predictions in the context of AI and work: Forecasters misread the cognitive tipping point (CTP); the tech industry oversells AI; and we all overestimate a business' ability to embrace the new operating models that AI requires. Let's dig into these.

Missing The Cognitive Tipping Point

Many analysts base forecasts on the S-curve. It's commonly used (and we use it in our model) to forecast technology adoption. The S-curve describes the life of a technology from creation to irrelevance. It's flatter at the beginning, steeper in the middle, and flat at the end. Picking the point on the S-curve where adoption of an innovation goes exponential is the most important driver of predictions.

Here's a famous example: eCommerce forecasts drove the stock market to record heights in the late '90s, only to crash soon after. Did eCommerce transform the retail industry as predicted? The answer today is obvious: It did. Traditional retailers now speak of being "Amazoned." There's no argument with the result, but in all the excitement of the internet boom, the point of exponential adaption, the tipping point, was predicted 10 years too early.

The S-curve for AI adoption includes two CTPs, one for consumer adoption and the second for business adoption (see Figure 5-1). The CTP for consumers is closer. AI, in a light form, is here today. Alexa is in our homes, while Amazon and Google are using AI to target ads. But the CTP for business is in the 2024 timeframe, much further out. Automation deficits begin to scale at that time. In short, not many forecasts have made or factored in the distinction between the use of AI in consumer applications like Netflix and its use in business applications, such as replacing underwriters in insurance.

But here's the big thing. We understate the change management headwinds and technical debt that companies face when adopting new technology. Your most recent American Airlines ticket was produced by the original Sabre System, designed in the early 1960s with a pre-IBM 360 operating system. General Electric (GE) still has more than 400 legacy systems in just two divisions.[74] These systems, and the surrounding culture, are a barrier to AI progression and job deficits. Businesses desperately want the digital advances that AI can provide but are mired in old systems and ways of operating.

Another challenge with forecasts is the assumption that we can predict human behavior. We're all a bit crazy, emotional, and unpredictable. This is why chatbots find us so difficult to understand. The noted economist Richard Thaler earned a Nobel prize for correcting the faulty assumption that underpins all free market economic theory — that people behave rationally.[75] Advancements in AI won't just appear and reconfigure the employment landscape; they must adapt to the existing conditions to fight deeply rooted cultural forces and human unpredictability.

Here are two simple examples. ATMs were supposed to eliminate bank tellers, just wipe them out, but it didn't happen. No one could see that ATMs would lower the cost to run a branch and lead to branch expansion, and guess what? A lot more tellers remain than anyone anticipated. Railroad expansion in the 19th-century US was sure to destroy the horse economy, according to many. But not so fast. Railroads maintained the need for horses for many years. People still needed to get to and from train stations. Horses had to wait for the automobile for real retirement.

Figure 5-1 More Forecasts Pick The Cognitive Tipping Point For AI Too Early

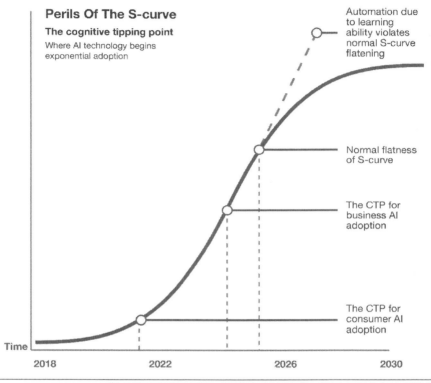

Source: Forrester Research, Inc. Unauthorized reproduction, citation, or distribution prohibited.

Chatbot enthusiasm combines all our worst prediction tendencies. Is there any doubt that text and voice input are the future human-to-machine interaction? This was clear to Gene Roddenberry, writing early scripts for *Star Trek* in the 1960s. And we're making progress. Growing numbers of Millennials use text and speech input for their phones. Why use drop-down menus, search screens, and clever navigations or download a clumsy mobile app to get what you need? Millennials simply demand the text option to ask a question and get an answer without precise spelling and construction. Hint: Older generations prefer it, too!

The future of chatbots is secure, but today, they're OK for only the most casual and simplest customer interactions. They're not ready for prime time. Chatbots for major banks, despite enticing names, only answer simple questions. They move the customer to a human or to a secure link on a website to complete a transaction. Why the slow progress? Chatbots aren't sophisticated enough to hold up their end of a social relationship. They can be boring, lose context over time, repeat themselves, or respond to keywords only. Let's face it: Consumers today just don't have patience, and major brands can't handle the risk.

If we had Dr. Turing looking over our shoulders, he'd just say we're not there yet. Alan Turing developed his famous test in 1950 to determine a machine's ability to exhibit intelligent behavior that's equivalent to that of a human.[76] Most chatbots use programmed rules, like today's interactive voice response systems, or search the internet or a knowledge base to provide answers, which is what Alexa does. AI chatbots, to be effective, must parse grammatically ambiguous statements to adapt to a conversation and grow smarter over time. Few do this today — and that's the difference between "lightning" and a "lightning bug" in the context of chatbots.[77]

Forecasting when AI will really kick in is one problem. Figuring out how fast and how far it will advance is another. AI has machine learning capability that will learn from new data and can self-adjust. It can revise history, much like our politicians and news media, to improve predictions, update classifications, and adjust patterns for inductive reasoning. Google search algorithms won't become obsolete; instead, they'll just get better. This means that, for AI, the upper flat segment of the S-curve may not apply or might have a lower slope. The S effectively becomes a J.

How We Built Our Model

We considered all these challenges in building our jobs forecast. The brains behind our model is J.P. Gownder, VP and principal analyst at Forrester, who leads future-of-work research at Forrester. Here, he describes his approach:

> We started with the BLS occupational database listing over 800 occupations, with job numbers for each. To understand the skills needed for each occupation, we went to the O*net database, which had a breakdown of the skills required for each job. We then looked at the Frey and Osborn progression of automation for each of these skills and assessed them based on Forrester's views of their likely progression.[78] As you can imagine, mashing all this together created a massive matrix and

model. But here's where we got creative. Based on our coverage of the 15 AI building blocks, we applied S-curves to automation progression for each occupation. This gave us the job dividends and deficits year by year. As we did in previous versions of our forecast, we included automation dividends — the jobs created by the automation economy. In this new forecast, it's much more granular, though: For each occupation, we devised ratios, i.e., for any deficit created, what would be the likely creation of new jobs? Finally, to make the forecast actionable and fit into the future-of-work framework, we mapped all this into 12 work personas, or generic work categories.

Our forecast shows that the AI tide will rise slowly but consistently as machines get more intelligent. They'll take more and more pieces of our current jobs. This approach follows patterns of previous general-purpose technologies. GPTs differ from single focus automations like Eli Whitney's cotton gin, which targeted textiles, or trains, which altered transportation. GPTs like the steam engine, electricity, phones, and the internet were slow and promiscuous. AI is the next one. It's already not a separate software market but a component of most technology products and services. It will affect all industries and every worker, task, and job.

6 AUTOMATION DEFICITS ARE INEVITABLE

Automation deficits are inevitable. There's no scenario where AI will add to the traditional workforce; we project a 16% reduction in jobs by 2030. Machines will just get better at doing the things we currently do. The big question is, which jobs will be affected, and how quickly will it happen? To answer this and help leaders and workers plan for the progression of control, scale, and convergence factors, we grouped the 800-plus occupations into 12 automation personas. Based on skills, tasks, physical requirements, and other factors, all occupations fit into these 12 categories. We also show existing jobs, the number of occupations, and automation status (see Figure 6-1).

For jobs within each persona, the tasks, skills, thinking, and physical actions are similar — at least, similar enough to be able to forecast the effects of advancing automation. Furthermore, based on a machine's ability to overtake the tasks performed, the timing of those effects will be the same. This simple classification of workers gives us a new way to think about jobs and allows businesses, governments, and individuals to plan better.

Here's an easier way to think about it: Let's use water as a metaphor for the rising intelligence of automation (see Figure 6-2). Employees of a certain persona occupy a specific floor in a building. The knowledge workers and executives might be on higher floors with better views. Cubicle workers and coordinators will be on lower floors.

Meanwhile, outside this building, automation is getting smarter, rising like a slow tide. Each month, the tide rises a little higher as automation improves in two important dimensions: 1) understanding context and 2) handling more variability. As automation improves in these two areas, it takes on more tasks that humans currently perform. The forces of automation chip away at a search here, a calculation there, a call no longer necessary, or a decision that once required thought. Workers first find a toe in the water, then a foot, then maybe an arm. The water rises. And automation deficits climb along with it.

Jobs on lower floors get flooded first. These jobs are machine-centric, meaning advancing automation can easily do what they do. Cubicle workers, offshore labor, coordinators who carry a clipboard, digital outcasts, and others on the wrong side of the skills divide will slowly disappear. Workers in these jobs will become deficits sooner than those on higher floors. These people are already wet.

Jobs on other floors are halfway in, or just above, the water line. Some knowledge workers fall into this range. Jobs that leverage a single domain of knowledge are more at risk for transformation or replacement. Function-specific workers like underwriters and loan officers are in this middle grouping.

Figure 6-1 Twelve Personas Capture Today's 800 Occupations

Trend		Number of occupations	Number of US workers
↑	Human-touch workers	76	15,196,250
↑	Cross-domain knowledge workers	92	9,668,290
↑	Teachers/explainers	64	8,844,680
↑	Digital elites	20	4,000,720
↓	Single-domain knowledge workers	41	5,533,780
↓	Physical workers	288	32,271,040
↓	Function-specific knowledge workers	50	6,357,980
↓	Location-based workers	70	29,937,620
↓	Coordinators	36	9,997,250
↓	Cubicle workers	72	20,676,290
↑	Mission-based workers*		
↑	Digital outcasts*		

Source: US Census Bureau annual projections for the US up to the year 2030 and US Congressional Budget Office labor-force participation-rate projections for the US up to the year 2030
*Mission-based workers and digital outcasts are both components of the category "evacuees" — that is, those who have exited the typical jobs economy. These workers will emerge from all 800-plus occupations.

Jobs on other floors are halfway in, or just above, the water line. Some knowledge workers fall into this range. Jobs that leverage a single domain of knowledge are more at risk for transformation or replacement. Function-specific workers like underwriters and loan officers are in this middle grouping.

Yet many jobs will stay well above the water line. Human-touch workers, whose in-person human presence is critical, will expand in roles and numbers. Workers to perform personal tasks that require intuition, empathy, and touching will be in great demand. These activities entail a level of variability that machines won't be able to respond to effectively. In many transactions, people will soon pay more to interact with humans rather than machines.

Most physical workers will remain above the water. For the foreseeable future, robots generally lack the physical agility to replace them. Cross-domain knowledge workers also need not shop for wetsuits. The connections they make cross too many information domains for machines to handle. We'll explore each of these personas with worker and company interviews to describe the characteristics of each.

Figure 6-2 The Water Rises As Automation Gains In Intelligence

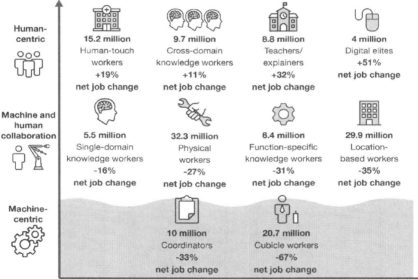

The water rises as automation handles greater context and variability

Note: Numbers of workers are totals for today's workforce. The percent automated includes the deficits or jobs removed from the workplace by 2030. Automation dividends will offset these percentages.
Source: US Census Bureau annual projections for the US up to the year 2030 and US Congressional Budget Office labor-force participation-rate projections for the US up to the year 2030; automation percentages are Forrester forecasts.

The Underwater Crowd And The First To Go

The biggest deficits, and the earliest, will occur in the cubicle worker personas, so let's start there. They include jobs in finance, accounting, procurement, and contact centers, among others. Software robots — both chatbots and RPA bots — are replacing these jobs today. For some, it can't be soon enough. Julie works in an accounts payable department and is ready to escape her cubicle:

> I've spent more years than I want to admit in a cubicle, and over those years I've learned that a cubicle is one of the worst things to wish on a person. Simply put, people weren't meant to be put in cubes. Now I know that might be the most efficient of ways to have people work, it's just not for me. Why on earth are you told to think outside the box, when you're in one all blasted day? They measure me 12 ways from Sunday. And this has been going on for some time. But now the focus is on figuring out what we do that can be done by a robot. And now I have to train the thing to do all my account updates.

Here's what Julie is telling us: First, she needs a new job. She feels that cubicle workers like her have been under attack from automation for decades, but new automation approaches have stepped up the game. These are the invisible robots, built with RPA or chatbot platforms, that are inching into her cubicle. Julie's work involves data entry, switching among applications, creating messages and reports, cutting and pasting, and simple calculations. It's work she finds mind-numbingly boring and repetitive, but she also has some deep animus toward the bots replacing her. And she should. Her tasks are ones that these invisible robots do well. Simple economics are at work. One bot that may cost $15,000 a year can offset two to four Julies. Routine work that remains in customer service and general office staff is now at risk.

Cubicle jobs like Julie's are on the lower floors of our building. Here's why: The automation technology to replace her job is here today. It doesn't require the "moonshot" approach that many AI projects on higher floors do. Medical applications, for example, must connect thousands of variables to make a diagnosis. This requires teams of data scientists, subject-matter experts, and vast computing resources. A simple software bot, by contrast, can be built in weeks and can replace well-structured repetitive tasks. As a result, cubicles are emptying humans out now. Today, the US workforce boasts approximately 20.7 million cubicle dwellers.[79] It's difficult to explain robots to these workers in an honest way. Here's Jim, an executive from a major health insurer, who tried to do it:

50

It seemed easy. I just sat down with them, the main guys. I explained how robots worked. That they were not physical — no arms or legs — but software that ran on servers and took over desktop tasks. They would update the reports for them, check those lists that needed to be verified, and send emails attaching the right documents. I told them the robots were going to be great. They will only do the things you hate to do. They will free you to spend more time with customers or thinking about ideas to make your job better.

This made sense to me. I truly believed it. But then they all went back to work and all I heard around the office was that these robots were going to take their jobs. This changed my way of thinking. I started to look at the robots from their perspective. And it's not good. Employees assume management has the worst intentions. It's not their first rodeo. They have seen the top floor come up with some BS story before, but at the end of the day, they just made life a bit more miserable and saved money somehow. We need to think this automation thing through. We need a better story, but that means understanding how things will end for our workers. And nobody really knows. Not management and not the workers.

Is Jim a bad guy? No. Does he take joy in eliminating reliable paychecks for his working-class and high-school-educated workers? Nonsense. What he lacks, though, is a view of the future. If he had one, it would show that only a small percentage of his workers will transition to the automation dividends we describe in chapter 7. He'd understand why workers are skeptical of robots and don't trust management to give them the straight story. He'd find a better way to communicate with workers and would invest in change management to guide the humans who keep their jobs.

Are you any safer in your cubicle in Bengaluru, India? Probably not. Despite the utopian workplace of fountains, open spaces, clean water, and smart young people on the move all around you, you may be feeling anxious. You look around and see hundreds of cubicles where workers answer calls or enter transaction data for global banks. But you know there's another story outside the heavily guarded gates, where the informal economy — 40% of India's workforce — is at full throttle. People huddle in tents, cook on makeshift stoves, and scramble. You count your blessings but realize that many of the good cubical jobs that have lifted India's middle class may soon join this informal world.

Here's why: The tasks that enterprises outsourced to offshore locations are bottom-of-the-pyramid work. This work is repetitive, structured, and well documented. And, no surprise, it's perfect for robots. Virtual agents or chatbots can now carry out this work at lower cost.[80] Rather than turn Harry, who works down the hall, into a deficit, managers will look first to offshore contracts. They've never met those people.

The result? Emerging automation will turn an industry used to double-digit growth into one with strategic issues.[81] Growth in revenue at the beginning of this century was 60% to 70% for five leading Indian business process outsourcing (BPO) companies: Cognizant, HCL, Infosys, TCS, and Wipro.[82] They're now lucky to show growth in low single digits. The good news? The average revenue per employee, a measure of their growing skills, is showing a positive upward trend.

This doesn't mean the end of major Indian-based and other offshore tech companies. They sit quietly, but powerfully, behind too many systems we depend on every day. Drive through that self-service toll booth, use online banking, or book a reservation for a hotel room — the contractor that built and maintains that service is probably Cognizant, Tata Consulting Group, or Wipro. And even if it was Accenture, a US-headquartered consulting company, it probably used what used to be referred to as "captive" talent in India. What are these firms doing to deal with slow growth? They're upgrading their AI skills.

Coordinators are another category of job occupation that will be underwater soon. Thirty-six occupations, with almost 10 million people, manage the work efforts of others. As we discussed in chapter 2, automation will soon maintain schedules, coordinate orders, and manage shipment logistics, the core of many coordinator jobs. Algorithms and software robotics will replace them.

Function-specific knowledge workers is another category soon to be underwater. Insurance underwriters (125,000 US jobs) and loan officers (350,000 US jobs) are examples for this segment, which contains 50 occupations.[83] What do they do? They comb through third-party data, navigate brittle in-house decision-support systems, and prepare reports. On average, these workers have college degrees and make $60,000 a year. But that salary, and the positions, won't hold up.

If machines had personalities and could talk, they'd tell you that their favorite job is correlation, and the more variables, the more fun. They'd view the work of today's loan officers to determine a credit score as mere child's play. They'll do this work, faster, with fewer errors, and at lower cost. And because they love data so much, they'd add sources like Facebook, Google Maps, and other social media to advance their analysis. They'd see you've had the same email address for five years and are on LinkedIn and are therefore a good credit risk.

The tech community will target these function-specific use cases. Enova Decisions, for example, has 50-plus analytics experts who maintain a machine learning platform called Colossus that's hosted in the Amazon cloud. It estimates that automation will replace 50% of today's white-collar loan offices and insurance underwriters in five years.[84]

In some cases, function-specific knowledge workers will become deficits, despite what may start out as good intentions by management. A midsize mortgage bank, for example, now has 36 fewer people in the loan department, but that wasn't the original plan.[85] Software robots now order tests, fill out disclosure agreements, collect data for risk assessment, and enter data into a cryptic loan origination system. A machine learning algorithm then creates a credit score. The people previously doing these tasks were mostly high-school graduates. They had reliable paychecks and health benefits, but no longer. The bank had hoped, and told employees who helped with robot design, that additional loan volume would allow it to handle more loans, meaning no layoffs. But when that volume didn't materialize, the bank had to let the people go.

These three categories — cubicle workers, coordinators, and function-specific knowledge workers — will produce significant deficits. Some percentage of these will transition to better jobs, motivated by higher wages and job security. They'll feel a slow burn that pushes them outside their comfort zone and have what we call constructive ambition, which makes you eager to learn. But many won't. We term these workers "digital outcasts," and they'll find no place in an automated world, unable to work effectively with machines due to issues with skills, attitudes, and ambitions, like Enzo, a machinist:[86]

> A few years ago, I retired after 20 years as a machinist. I
>
> was worn out. They worked me hard. I then started to
>
> drive a cab for a living. But then I saw a new machine

shop was hiring a mile and half from my home in Topsfield, Massachusetts. It's good to see machine work coming back to the US after so many jobs lost.

I'm no rookie. At 56 years old, I managed to get hired. I had changed some — added a few extra pounds — but the machines had changed a lot more. They had gotten very high-tech. The work had changed. There was now a lot more to do in the prep — setting up the machine for the job — putting in codes — programming. A kid getting his college degree tried to help me. He explained the setup and the testing process. But I was just not interested in learning the new machines. I hated to come in every morning. Putting in the codes scared me. I thought I would break something. I had been a top performer for decades at a top shop. Here, I was a slacker.

So I went to the foreman and resigned. He tried to talk me out of it, said he liked me. That I kept to myself and worked hard, wasn't pretentious. I said it was too hard for me to focus on something I wasn't interested in or good at, and I asked him, "Can't I just run the machine?" But the foremen shook his head and said no. We need someone end to end. So I quit. I went back to my cab. It's relaxing for me and I don't need much. I take care of my family and stay out of trouble. That's what a good paisano does.

Here's what Enzo teaches us: There are many in today's workforce that don't have the aptitude or attitude to transition to work with advancing machines.

Workers will need to develop constructive ambition, and employers must learn to test for it. Transformation will require workers to go beyond their normal patterns. Like the early hunter who follows a deer deep into a new forest, they must be willing to encounter a different reality.

Appalachia: The Worst-Case Scenario For Digital Outcasts

West Virginia alone boasted 140,000 coal miner jobs in 1940; as of 1917, there were only about 11,600 left.[87] The coal jobs aren't coming back, and there's little to replace them. Worker discontent is at an all-time high, particularly for white men lacking college degrees, which is a large percentage of the population in Appalachia. It's hard to find a more challenging place for workers to adapt to the digital economy. The Amazon team that searched for a new headquarters location never gave it a thought. Here's a firsthand view from Doris Trammel:

> I'm a courtroom adjunct in Eastern Kentucky. I see them come through every day almost. They are 25 years old but look 40. They stare more at their shoes then the judge. Their job is distributing opioids and somehow, they got caught. The Eastern District of Kentucky where my court is at is ground zero. Painkiller and heroin abuse is rampant.[88] They didn't wake up some day as a kid and say, "I really want to manage a drug supply chain." No, they grew into it. They wanted the things they saw on TV — the flat-screen TVs — the fast cars — the nice clothes. They dropped out of high school and looked around and saw no way to get those things, no jobs for the unskilled and poorly educated. The unemployment rate may be low right now in this country but it's pretty high here. We have thousands of folks who have just given up looking. Most have somehow connived to get some form of public assistance. And they are good at it. Their parents trained them.

Employment is fundamental to a stable and productive community. Without it, youth in rural communities will likely drift into crime and addiction. So how will the youth fare in the future of work? If anyone could explain this, it would be Nick Mullins, a ninth-generation Appalachian and the fifth generation to work in the coal mines. He's also the first in his family to go beyond high school. He ended up working in the mines for five years but left; he wouldn't subject his family to the dangerous water quality and mining practices. Nick tells us his story:

> What's it like in Appalachia now? Here's what we have: scarred mountain tops, toxic slurry ponds, contaminated water, and local populations that are depleting as fast as the major underground coal seams. Visitors enjoy towns with museums and exhibition coal mines like Lynch and Benhams, Kentucky. But tourism only helps a few local businesses and doesn't provide the living-wage jobs we need. Few tourists learn about the other histories of Appalachia. They don't see places like the Hurricane Creek miner's memorial near Hyden, where 38 miners lost their lives the year after the 1969 Coal Mine Safety and Health Act was passed.

> Most of the guys I worked with in the mine were damn good people, even some of the foremen. They worked hard, and most had families that loved them — families they'd do anything for. We called Johnny Vanover "Old Man" because he wouldn't let us call him anything else. He told me he started in the mines at age 18. I think by the time he was 70 he'd spent over 50 years working underground. He never failed to show up, and often worked 12 hours a day, six days a week. Many of us would ask him why he didn't retire. He'd say, "I don't

know what else I'd do with myself. I enjoy it," but that was only half of the story. He had paid to send all his children to college and was then paying to send his granddaughter to college.

It was hard and dangerous work, though. We all lived in constant danger of roof falls, electrocution, crushing injuries from equipment, suffocation from oxygen deficient air, or explosions—not to mention long-term illnesses such as black lung and silicosis. There's also very little job security. Large coal corporations, doing as they often do, shut down most of their mines when the markets are down. A few file for bankruptcy, with executives stripping employee pensions and retirement healthcare benefits.

Miners must find another work life, but the options are severely limited. Boom-and-bust markets have always worked to the advantage of coal companies, while leaving Appalachian communities to suffer. In many of our coal-producing counties, over 90% of the mineral rights are owned by absentee owners. When markets are up, coal companies open mines and extract coal as fast as possible. When the markets are down, they idle or close their mines, file bankruptcy to get out of mine reclamation costs and debts to their employees. Meanwhile, high unemployment rates contribute to deep poverty in mountain communities, with all its many symptoms: substandard housing, depression, and substance abuse.

People say, "Well, why don't you just leave?" Many have with each bust in the coal markets. There are a few country songs about all the people headed to the big cities for work, but now it's hard to find good jobs in the cities for people with only a high school diploma. And many people can't leave. When the coal market is down, so are the housing markets. Getting pro-coal politicians to build infrastructure for alternative economies is extremely difficult as well.

We have poor internet services in the mountains. We have some DSL and some cable internet near the towns. But that's not going to plug us into the new economy. What work was done in the 1990s to provide network infrastructure in Virginia's coalfield region only brought in opportunistic companies seeking tax breaks and cheap, desperate labor. Travelocity was one of them. I worked in a call center for seven years living paycheck to paycheck before I decided to go into the mines in search of better pay and benefits for my family.

What other service jobs are available do not pay nearly as well or give you the same feeling of sacrifice and community as working in the mines. If such jobs paid as well as coal mining, I'm sure miners would adapt. But since there's no hope for something better, the sacrifice of working in the mine is worth the risk, and miners come to enjoy the sense of pride and community from working in a dangerous place.

Here's what Nick is telling us: It's hard for many miners to see themselves in roles like working in a call center or as a technical guy or programmer. We need to understand this. People from outside the region, especially academics, politicians, and government agencies, tend to come in and try to tell residents what to do and how to do it, as if they're too ignorant to know. Few of them have faced the hopelessness of living in an extractive mono-economy, as Nick Mullins has. This lack of experience shows when they come up with ideas and programs that don't work. Unless they can help miners pivot their skills in a new direction that provides a long-term, living-wage job alternative to coal, they just come across as more liberal elitists from up North. Planning for the future must be a collaborative discussion that includes local people's viewpoints, talents, and dreams.

Hovering Just Above The Water Line For Now

Some occupations sit just above the water line. They're close to being disrupted by automation. These require less physical agility, have narrower sets of knowledge that machines can tackle, and require less human interaction. The segment has two forms of knowledge workers and includes location-based workers in retail who will see jobs transformed through digital and physical convergence.

We expect that single-domain knowledge workers will get a lot of help from machines. We estimate that 41 occupations with 5.5 million jobs fall into this category. Here, decisions are confined to a deep but narrow set of knowledge. Machines can make these connections, and variables are known, constrained, updated, and repeatable. A radiologist is a good example. Computer vision and machine-based data correlations can reduce the number of radiologists, detecting tumors in MRIs far faster than humans can and with similar and ever-improving accuracy.[89] Actuaries are exceptional at math but deal in a single domain that a machine can conquer.

Single-domain knowledge workers are working in general office environments. But some 29.9 million workers across 70 occupations are tied to a physical location. Location-based workers must be in a certain place, such as a retail store selling shoes, at a ball game being the umpire, or in an office building lobby providing security. The physical environment defines their jobs now, but as the water rises, convergence — which embeds digital smarts in physical locations — will define them.

Four Segments Stay Well Above The Water Line

Automation is getting smarter and taking on more things that humans do. But some occupations put more demands on automation than others. Human-touch workers, cross-domain knowledge workers, teachers and explainers, and physical workers will remain above the water line for now. Let's start with the human-touch persona, some 15 million jobs across 76 occupations.

Human-touch workers are nurses, physical therapists, the "companion" taking care of your mom, and even funeral directors. Intuition, empathy, touching, and physical and mental agility are all essential, and they're too much of a challenge for machines today. Machines are nowhere near capable of performing most of the required tasks. Rapid and real-time reaction to changing human contexts makes this type of job a super challenge for machines. Though robotics firms in Japan are developing robots in these categories to care for an increasingly aging population, the fine motor skills required will take a decade or more to master. The empathy will take even longer.

Cross-domain knowledge workers are also a challenge for machines. These jobs require intuitive connections across diverse information domains. Machines fail when a decision or task requires out-of-the-box connections — meaning out of the machine's defined data model — particularly when the domains shift over time. Emergency room physicians, lawyers, and marketing professionals who build product strategies, for example, are safe; they're among the 92 occupations employing more than 9.7 million people in the US today. There's just too much variation in the tasks for a machine to replace them.

But perhaps the most misunderstood category in the automation debate is physical workers. Reading media coverage, you might think an army of robots, metal heads glinting in the sunlight, will soon march into our factories and take our jobs. This isn't the case. Automation deficits have been oversold for the 288 physical worker jobs (roughly one-third of all occupations). These workers perform physical activities that require arms and legs as well as body movements such as climbing; lifting; walking; stooping; and scaling ladders, scaffolds, or poles.

Here's why these occupations remain above the water line: Firstly, we can't produce enough robots fast enough to affect physical workers in a material way.[90] Current robot production levels could take away only 6.7 million jobs a year from 340 million physical workers worldwide.[91] And secondly, for most use cases, machines can't replace human agility at costs that make sense. You can train a human to put a bolt in a different spot in a second; the cost of programming this change into robots will be very high. Here's Prasad Akella, CEO of an AI startup called Drishti, who has spent a great deal of time on factory floors.[92] Prasad explains:

> In my graduate work I tried to build a robot arm to do things. It's amazingly hard to do the simplest human things. Get a robot to take a pen out of a pocket. Believe me, it's hard. I started working in factories and built the first collaborative robots ("cobots"), to amplify workers' physical capabilities and saw the struggles of automation firsthand. Despite advances in robotics and automation, most factory work is still performed by humans. Take that smartphone out of your pocket. How many robots helped build that? I'll tell you. A big zero. Probably around 100 humans, most likely in China, built that phone. Globally, only about 10% of factory tasks are performed by robots.[93] And here's what bothers me with all these forecasts: The people that do them have never been on a factory floor.

> Here's the takeaway: Machines will be hard-pressed to replace human agility at scale in the near term. Humans have phenomenal ability to handle physical variation. The cost of replicating this in robots will be very high. And even with accelerating progress, we lack the production capacity to materially affect the human workforce.

Preparing For The Flood

Automation will affect certain personas sooner and in greater numbers (see Figure 6-3). Each small square represents 100,000 workers affected in one of three ways. We show the automation deficits, dividends, and retained jobs in today's workforce. Projections by personas can help you develop specific plans for that group of workers in your organization.

Figure 6-3 Automation Will Result In A 16% Job Reduction By 2030

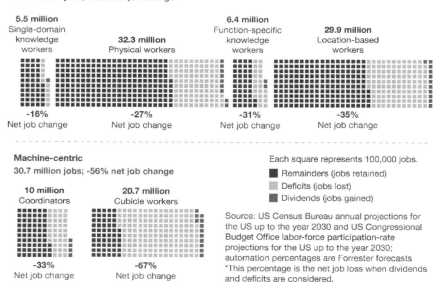

2030 employment will be reduced by 16%*

Deficits by 2030	Dividends by 2030
29%	13%

Human-centric
37.7 million jobs; +23% net job change

15.2 million Human-touch workers	**9.7 million** Cross-domain knowledge workers	**8.8 million** Teachers/ explainers	**4 million** Digital elites
+19% Net job change	**+11%** Net job change	**+32%** Net job change	**+51%** Net job change

Machine and human collaboration
74.1 million jobs; -30% net job change

5.5 million Single-domain knowledge workers	**32.3 million** Physical workers	**6.4 million** Function-specific knowledge workers	**29.9 million** Location-based workers
-16% Net job change	**-27%** Net job change	**-31%** Net job change	**-35%** Net job change

Machine-centric
30.7 million jobs; -56% net job change

10 million Coordinators	**20.7 million** Cubicle workers
-33% Net job change	**-67%** Net job change

Each square represents 100,000 jobs.
- ■ Remainders (jobs retained)
- ▨ Deficits (jobs lost)
- ■ Dividends (jobs gained)

Source: US Census Bureau annual projections for
the US up to the year 2030 and US Congressional
Budget Office labor-force participation-rate
projections for the US up to the year 2030;
automation percentages are Forrester forecasts
*This percentage is the net job loss when dividends
and deficits are considered.

7 AUTOMATION DIVIDENDS DON'T FILL THE EMPLOYMENT GAP

Automation dividends are the new tasks and jobs that result from automation. Many of the dividends are visible only after the automation is in place and, sometimes, years after an innovation is absorbed into the economy. The highest number of dividends will be in five personas: mission-based workers; human-touch workers; teachers and explainers; digital elites; and cross-domain knowledge workers (see Figure 7-1). This chapter touches on all five personas but drills deepest into digital elites and mission-based workers.

Digital Elites — The Rising Stars Of Automation

This persona includes tech industry executives, TED talkers, machine learning specialists, and many other digitally based occupations. Digital elites will see their wealth rise in comparison with the 11 other personas. The Facebook campus is full of them. The headquarters with the famous "thumbs up" sign in Menlo Park has free meals, dry cleaning, a gym, and even a barber shop, but not for contract workers.[94] Contractor Jiovanny Martinez, a security guard at Facebook, can't access these benefits. He also drives for Lyft and works as a park ranger to support his family. Maria Gonzalez, a janitor at Facebook, says, "It's not the same for janitors. We just leave with the check."

Big data specialists, process automation experts, information security analysts, user experience and human-machine interaction designers, robotics engineers, and blockchain specialists are just a few of the elite occupations. The elite of the elite will be machine learning experts. Like the algorithms they will master, they require nondeterministic thought patterns.

In this important way, they differ from today's top-performing programmers, who are likely to be better at calculus than at statistics. And they're in demand. Various estimates forecast jobs well into the millions right now for those with "analytics skills."[95]

Figure 7-1 Automation Dividends Reward Five Personas

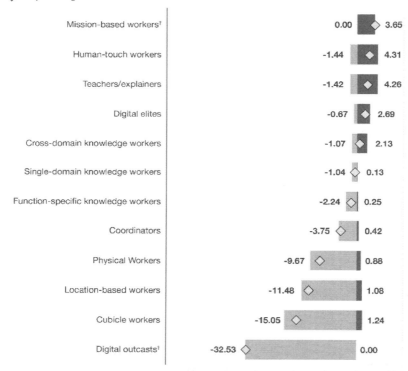

Job changes (in number of jobs) by 2030*
Millions of US jobs

- ■ Dividends (jobs gained)
- ▨ Deficits (jobs lost)
- ◇ Net job change

Mission-based workers[†]	0.00	3.65
Human-touch workers	-1.44	4.31
Teachers/explainers	-1.42	4.26
Digital elites	-0.67	2.69
Cross-domain knowledge workers	-1.07	2.13
Single-domain knowledge workers	-1.04	0.13
Function-specific knowledge workers	-2.24	0.25
Coordinators	-3.75	0.42
Physical Workers	-9.67	0.88
Location-based workers	-11.48	1.08
Cubicle workers	-15.05	1.24
Digital outcasts[†]	-32.53	0.00

*Forrester forecast
[†]Mission-based workers and digital outcasts are both components of the category "evacuees" — that is, those who have exited the typical jobs economy.

We'll see trickle-down dividends in the form of specialized and personalized services, and not just for the landscaper in the back of the pickup. The digital elite will need dog-walking services, personal organizers, and shoppers. *Secret Lives of the Super Rich*, the highly forgettable CNBC series, even highlighted people moving into these jobs. Closet organizers for the well-to-do, personal assistants, or specialist Ferrari shoppers are all job dividends, but there won't be enough of them to move the job recovery numbers any time soon.

Automation Curators Take Care Of Robot Monitoring And Training

Automation is expensive. Many jobs don't require the skills of digital elites. They're new work patterns that emerge when humans and machines work together. But what are some of these jobs? They include content curators who fall into the function-specific knowledge-worker persona; librarians who maintain machine learning training data sets; and various engineering, design, maintenance, support, and training jobs. When machines make decisions, for example, new teacher/explainer roles emerge to validate machine predictions, take on exceptions when the machine doesn't have enough data, and explain to people what's going on. Chief trust officers will help customers with blockchain-enabled currencies. Data detectives will unravel the mysteries of big data applications. A new breed of controllers will monitor highways and airways crowded with self-driving cars and drone deliveries.

We derive our forecast of dividends from the number of job deficits. We created a ratio for each persona relative to automation deficits. A heuristic used in RPA projects can make this clearer. If a software bot eliminates 100 humans from entering financial data into a general ledger, the corporate project holds back or budgets 10%, or 10 new jobs, to support the automation. These are higher-paid and more skilled employees than the 100 replaced. And what are they doing? Two-thirds are in various business segments designing and maintaining robot software, training humans, and dealing with change management. The remaining third are in a technology management group or IT. They'll control robot operations and manage the technology infrastructure.[96]

New field-service workers who keep the robots running will be a dividend as well. Multibillion-dollar investments in self-driving cars, for example, will quickly enable many forms of autonomous robots. These will roam around buildings, work as security officers, and provide a telepresence in office environments. And someone needs to fix them when they break — which they will. These dividends will be mainly in the physical worker and location-based worker personas, and there will be many. It's likely that enterprises like HP or Xerox will have "robot managed services" to augment their traditional server, desktop, and printer maintenance.

Our model estimates that by 2030, due to new automation roles, we'll see the creation of automation dividends of 13%. This means more than 20 million new jobs in the economy.[97] We touch on only a few examples here. For example, convergence will require a host of new design skills that combine the digital and physical worlds. Plenty of tech firms already have incredible software chops, but they lack physical design and basic engineering skills. Product design firms are already seeing an uptick in engagements with startups that need to augment their software teams with knowledge of physical design.

AI Assistants Are A Supporting Dividend Category

Many professionals today use their extensive education and experience to make decisions. They base these decisions on some form of prediction. As explained by economists in *Prediction Machines: The Simple Economics of Artificial Intelligence*, the real effect of AI will be to lower the costs of making decisions to near zero by moving them to machines.[98] We factored these trends into our forecast for the three knowledge-work personas and found something interesting.

As you disrupt cross-domain and single-domain knowledge workers, you get job dividends in other personas. Many of our most skilled professional jobs will be performed by a greater number of less trained people. Function-specific knowledge workers, coordinators, and human-touch workers can migrate to better-paid positions. Tech-savvy physician assistants and nurse practitioners, for example, will be able to do more under a machine's direction. Society will benefit from a lower cost of important services.

Teachers And Explainers Will Take On Elevated Importance

There are currently 64 occupations that fall into this persona, which includes more than 8.8 million jobs. It includes traditional teachers but adds thousands of new positions that require teaching skills. As machines take on more knowledge and more repeatable tasks, demand will grow for human skills to communicate, explain, and present information. And not all of these jobs will require a four-year degree. We'll see the creation of new roles to explain what machines have done, what they will do, and how they work.

Mission-Based Workers Are A Dividend As Well

Here's a story that may or may not be true: When John Lennon was 5 years old, his teachers asked him what he wanted to be when he grew up. He wrote "happy." They supposedly told him he didn't understand the assignment, and he supposedly told them they didn't understand life. Regardless of the veracity of this tale, a reset of life goals may be the biggest automation dividend we receive.

Already, we're beginning to see a shift in people's attitudes and behavior. Practicing meditation to reduce stress has become mainstream. In total, yoga, meditation, and mindfulness services are expected to grow to an $11 billion industry by 2020.[99] Today, 9.3 million Americans meditate, fueling a $1 billion-plus industry.[100]

People also just want to play more and work less. The Entertainment Software Association reports that 42% of Americans play video games for at least three hours every week. But here's a surprise: three-quarters of players are adults, not teens, and 44% are female.[101]

These trends suggest an approach to life that's more personal than work-centered. We created the mission-based worker persona to capture this trend. These people may toil from home via the internet, wearing sweatpants and a tee shirt. They may be overqualified, unemployed tech workers. They could be working class, college educated, or Baby Boomers, but they all share an important attitude: They look for meaning at work, a mission, something deep and passionate. If they can't find a job they want, they'll try to create their own.

These workers want to live in the present. A pension at age 65 is no longer a goal, if you could even find such a thing now. Some will be drawn to a version of the triple bottom line (social, environmental, and financial) or to certified B Corporations, which receive at least a minimum score for social and environmental performance.[102] They may avoid a corporation that shows a strong profit but owns coal production. Others may support the more conservative "freedom fighters" position on gun control. No single set of opinions or goals defines them. Mission-based workers seek values more than money.

Millennials will make up a strong percentage of mission-based workers, mainly because they're now a third of the US workforce.[103] They graduated into the Great Recession, entered a weak job market, and now face record amounts of student debt. Entering the real world, they may find that making an impact and climbing out of debt takes time and requires development of skills. They're ripe to select an alternate life style.

Mission-based workers are a worldwide trend. Japan uses the term "herbivore" to describe men who have no interest in getting married or finding full-time jobs.[104] A gradual decline of the economy has contributed to the rise of herbivores, who turn their backs on typical "masculine" and corporate roles. Japan's part-time employment has steadily increased over the past few decades and, as of 2017, comprised 22% of the workforce.[105] Youthful part-time workers are specifically called out in the Japanese workforce, and since the 1980s, when the Japanese economy was prosperous, have been known as "freeters."[106] Here's just one of a thousand stories we'll hear over the coming years from people like David White:

> I worked in investments in NYC but left that simply for
>
> an emotional need. I felt alone in NY, and as a

"momma's boy" moved back home. I stayed on top, however, with the New York experience and Vanderbilt education impressing everyone back in Atlanta. A boy of privilege. I continued in pension fund management and [parlayed] that into business development through the dot com bubble within the telecom field. Well appreciated, but sensing the work was fun but merely a game, my internal drumbeat to live for a greater cause helped me to slowly adjust my focus, remembering my early love for others not like me. With the support of my spouse, and after [the] birth of our first child, I entered social work, coupling my MBA with a new MSW in child welfare. The shock of a life of risk and attempt at deep meaning was in the paycheck, where the first one was 30% less than what I was used to. I had saved well and plugged away, searching for a place to truly live. I found the child in foster care. We adopted our second child [who was in foster care] from China, and I began to listen to children and their needs all the time. In 2010, I decided many more adults would love to help children in foster care, but don't know how. From that point, my life's purpose, and my internal drumbeat, became in sync. Fostering Great Ideas, ideas that reform foster care through increased community engagement, birthed.

The greater number of children entering foster care recently is a reflection on the stresses from addiction, and recently the startling opioids addiction, that has plagued all parts of our society. And I'm concerned with

the spike in foster kids — the number of kids in the
system has increased consecutively over the years.
Family stress used to be a matter of economic stress for
many. Now, addictions can ruin any best-laid plans,
throwing a wrench into an otherwise functioning family
unit. Coupled with the increasing loss of two-parent
families over the last 30 years, undue stress seems with
us for a long time. "When will the pain end," many are
asking.

Here's the takeaway: Jobs will no longer define a person's sense of
meaning, as in days past. People will adopt a less material and work-centered
lifestyle.[107] The David Whites of the world will accept a less reliable job and
income to feel good about themselves. The tiny house movement, capsule
wardrobes, a move out of urban areas, fewer possessions, and decluttering
are all trends that support an increase in mission-based workers.

Departing The "Stuff" Economy

Most people who become mission-based workers have a less material
approach to life. Nick Mullins left Appalachia for environmental reasons. But
here's what really bothered him:

This uncertainty and misery of a hopeless situation is
always right there — ready to creep into whatever
thoughts people are having. The consumerism that
everyone is subjected to doesn't help either. We have
tried to protect our kids from the piped-in unreality of
TV and social media where everything is done to excess.
That was not the Appalachian way. But you can only do
so much. It just seems we are barraged with marketing
for a bunch of crap we don't need. We are told — go
work your ass off so you can make what you need to buy
all this stuff. If you do, you are being an American —

contributing to society. This keeps the economy going and people working. But I realized this wasn't true. Most of the crap was being made in China anyway. It's time for something different, something that takes advantage of both worlds: the self-sufficiency that allowed our ancestors to survive for hundreds of years, and the new technological world where we use human advancement to become even more sustainable and perhaps — even happy.

Here's where Nick is going: Safety nets from governments are in decline. Employers don't offer the job security they once did. If you work in a cubicle, are a digital outcast, live in a declining mono-economy like Appalachia, or work for an offshore outsourcing firm, you may have to choose between material goals and life goals. This means you may opt to become a mission-based worker. Twenty years from now, we may look back and say the biggest dividend from automation has been our attitude toward work. In a sense, there may be only one way to approach the future of work, and that's from within.

8 THE MURKY MIDDLE CHALLENGES JOB TRANSFORMATION

At first, the primary technology groups wanted nothing to do with Deutsche Telekom's Sebastian Zeiss' robots. They said they probably wouldn't work and would run afoul of internal regulations. So, he built his own farm — a bank of servers running RPA bots. So far, the robots have transformed hundreds of jobs but replaced not one. Sebastian tells it this way:

> The robots on our farm are here today, not something you read about that may happen. They don't have arms and legs or move around making strange beeping sounds. They are quiet and hidden. People are disappointed when we show them the bank of servers. There is not much to see but the robots are hard workers. They do 35Mio transactions a year, taking tasks that people didn't enjoy [doing]. They gather information on order status, send notifications, do calculations, and make simple decisions.

> Initially, the workers council was not happy. First reaction? Management always has a story about making our jobs better but what they want to do, what they need to do, is reduce costs. These robots are just the beginning. They will grow smarter and take our jobs. They prepared for a big fight. So, we entered a formal

agreement, that there would be no layoffs. No one would lose a job because of automation. We would retrain people for other more interesting jobs and use the saved hours to re-allocate work. Then it got better. We mainly automated tasks that people didn't enjoy at all and a lot of employees considered the bots as a partner rather than an enemy. We decided to call the bots [front-end assistants] instead of bots. That also helped as it gave the notion of personal supports instead of something that takes away work.

Here's the takeaway: For many enterprises, the initial thought is to restructure existing work, not to eliminate people.[108] Job displacement and job transformation will occur side by side. Here's some data to support Sebastian's experience. Forrester asked 300 company directors across the US, Germany, Switzerland, and Australia to gauge the effects of software robotics on their employees.[109] The top answer? "Software robots will take over routine tasks that allow workers to focus on more strategic work." The second major effect was raising employee morale. And the third was to allow more customer communication. Dead last was "to reduce the need for human employees." George is an executive for one of the major credit unions and looks at automation this way:

Automation — and particularly AI — will allow a more human experience for our members. And when you think about it, that's all we have. We don't have the scale, the best online banking, or the number of services the big guys have. We must fight them with better customer engagement.

AI can help to merge our internal data with outside customer data and sift through interactions they have with us online, over voice, and via other channels. We need to figure out what a customer is trying to do and add a personal touch. In some ways we want to return to that bank experience from 50 years ago when you walked in, the banker knew you, maybe went to school with a cousin, knew your uncle, and gave you credit for being who you are.

Here's where George is going: We can't return to the old pre-scale days. They're long gone. But here's what we can do: give employees a few more hours to be a bit more human and try not to replace them.

Physical Workers Will Get A Needed Lift

As we saw in the chapter on deficits, occupations in the physical worker persona are above the water line. Robots aren't yet agile and cost-effective enough to replace them. Machines can beat a grand champion at chess, but walking 5 feet is a challenge for them.[110] As a result, transformation and augmentation of physical jobs will be more common than replacement.

Tasks that must be performed in challenging physical conditions will be augmented with robots for safety and health reasons. The cobot (from "collaborative robot") market, to pick one segment, is expected to grow from $400 million annually today to $9 billion in 2025.[111] In other words, the market for robots working with people is exploding. Exoskeletons are another example. These are full-body suits or vests that give the wearer physical support and added strength but aren't restrictive. They recall images of Iron Man and have a simple goal: to combine a human's cognitive ability with a machine's strength. There are hundreds of physical environments where tasks don't repeat enough to cost-effectively build a full robot with the algorithms, sensors, or agility to replace a human. Construction jobs, for example, often require lifting heavy items from one floor to the next as a one-off activity. A robot couldn't renovate the bathroom, but an exoskeleton can help carry that new shower stall up two flights of stairs.[112]

Long-Haul Truck Drivers May End Up In Cubicles

A former truck driver in Utah comes into the office and goes to his cubicle. But it's not a regular cube. He sits in what looks like a truck driver's seat. 3D images surround him. He's now driving around Cleveland using a combination of augmented reality (AR) and virtual reality (VR). He's not dealing with just a single windshield view, with an occasional glance at the speedometer. There are now 50 views demanding his attention. This is one vision of the future. In the shorter term, a long-haul driver will support a new self-driving upgrade and become an "in-vehicle systems manager." He'll oversee speed, braking, and steering. He'll monitor a dashboard filled with diagnostics, optimize the route, and take control for the final 10 miles to navigate aging loading docks. These workers now interact with machines in very different ways. The jobs require a different personality and new skills and attitudes. The workers must rebuild trust and control from scratch.

Knowledge Workers Will Experience Edge Transformation And Anxiety

At the top of the worker food chain are our knowledge workers. In chapter 6, we broke them into three categories. Of the three, cross-domain knowledge workers will see the least impact of automation in the next 10 years. They make critical decisions that require making connections that machines will find difficult. Choosing an acquisition target or a legal strategy, for example, will remain distinctly human. Single-domain knowledge workers, such as actuaries or radiologists, will experience more job disruption, while function-specific knowledge workers, such as insurance underwriters, will experience the most. But all will struggle with anxiety due to loss of autonomy and control. The roles below them have less control to lose. Cubicle workers, for example, are surrounded by apps and automation that lock them down. But all three knowledge positions will lose well-paid support positions that surround them. Nichole Jordon is a managing partner at Grant Thornton, an accounting firm. She explains the new machine- versus human-driven approach to investment:

> In my world an analyst might keep track of 30 acquisition targets. Data about financials, patent filings, negative citations, litigation, customer feedback, and culture all needs to be assessed. We now use AI to do the heavy lifting. Bots scrape data from websites that are brought into an AI system for analysis. Dashboards are automatically created. The machine stops short of recommendations to feed the acquisition pipeline but the data gathering, the sorting and modeling of the information is amazing. And we can bring in a lot more data — such as employee feedback from Glassdoor and other social media posts. This data gathering was human-driven before and now its machine-driven.

Here's another example: Goldman Sachs uses a machine learning platform to replace a human-driven reporting process, not a midlevel operational report but one prepared by top economists.[113] Here's how it works: The machine scrapes data from the US Department of Labor website. Based on past responses to similar employment changes, it predicts stock performance and sends the prediction to hundreds of traders. A team of economists and analysts used to do this, but the machine now completes the task within minutes after the data arrives.

A Centaur Transformation Will Release Much-Needed Human Talent

Designing employee experiences for the world of machines is of small concern today. But over the next decade, it will become a central issue. The centaur from Greek mythology is half man and half horse, and this concept can help us divide work between machines and humans. Think of the human as the head and of machines as the back end (see Figure 8-1). The middle section is where humans and machines must collaborate.

Figure 8-1 A Centaur Approach To Job Transformation Unleashes Human Talent

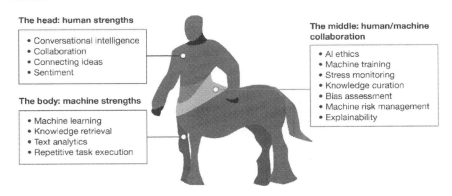

The head: human strengths
- Conversational intelligence
- Collaboration
- Connecting ideas
- Sentiment

The body: machine strengths
- Machine learning
- Knowledge retrieval
- Text analytics
- Repetitive task execution

The middle: human/machine collaboration
- AI ethics
- Machine training
- Stress monitoring
- Knowledge curation
- Bias assessment
- Machine risk management
- Explainability

The Head Becomes The Critical Asset

The centaur's head is human. Today, machines can't compete with a human's ability to communicate with other humans. But why would we expect anything different? We evolved to communicate with each other, with verbal and nonverbal nuances. Early humans who couldn't distinguish a friend from a foe during an encounter didn't last very long. Despite Alexa in your kitchen, it's impossible today to program this human talent into a machine. Here's Sally from a major telecommunications call center:

What happens when chatbots or virtual agents start taking customer calls? I'll tell you what happens [laugh]. The calls that get through to us become really hard. You think a stupid bot searching through some lame database can handle the crazy conversations we do. I had this confused person the other day. She wanted to tell me all about how her sister-in-law somehow screwed up the account — way too much information. The bot would have no chance. These new tools take the easy ones that we used to get credit for. I look around, my coworkers and I have a little more education each month, and the average call times are just exploding.

I had a drink with my manager the other day. A good guy. Not real sharp but a straight shooter. His boss is the head of operations, who is in deep guano. He said the new chatbots would reduce contact center employees by 15%. But that's not what's happening at all. Yes. The number of calls to the center has dropped some but the calls getting through are killers. So, we have to hire more people.

Here's where we are today: We're not quite ready to unleash chatbots on customers. They're unable to carry a conversation and, today, are likely to create bland, confusing, and irritating exchanges. We need to think of new ways that humans can provide memorable customer experiences. The most important part of the restructuring of work is to lift the human experience for both employees and the customers they serve. For many companies, future margin growth will be in the memories they create — and humans will be creating them.

The Back End Manages The Knowledge And The Heavy Lifting

The back of the centaur is the machine's domain. If machines could somehow express feelings, they'd say they're happiest crunching numbers. Connecting variables, computing probabilities, and searching knowledge domains are their specialties. Task harvesting — using software robots and machine learning algorithms to eliminate routine work — is a key part of the centaur's back end. It's doing the heavy lifting. And what happens when we give these tasks to a machine? Humans can become more creative and develop talents that we've stripped from them with narrow education and jobs like cashiers or invoice approvers. A 5-year-old child can find 10 ways to build something out of a pile of rocks. In most people, where's this creativity 20 years later? The more we automate the lower-level tasks, the more we'll upskill humans to unleash the creative instinct.

Enterprises can take advantage of this shift. They can move staff from the back office to the front. Instead of tasks that no one will see and few care about, new tasks that help customers will emerge. Here's an example: Sadly, some hospitals have more administrative staff in billing than they have nurses taking care of patients. Billions are spent in the US each year on administrative work to support a three-party (provider, insurer, and consumer) payment process that returns nothing of value to patients. But software robots will help. Texas Children's Hospital in Houston, one of the top children's hospitals in the nation, used software robots to help claims examiners.[114] Cognizant, an IT services firm, built them. The robots now process hundreds of claims a day. As a result, maybe the hospital can hire more nurses.

Many tasks we ask workers to do are just boring. Home furnishings titan IKEA foresees a future where robots eliminate many of these tasks.[115] IKEA's human resource center in Sweden receives 8,000 new employment contracts per year, and four full-time employees enter data in 50 fields for each contract. Software "robots" will replace these cubicle dwellers, and for a senior operations manager at IKEA in Sweden, that day can't come soon enough:

> Today, we lose good people who don't want these super boring tasks. To do it all day long is simply torture. We're not afraid of robots because we see machines supporting workers, not removing them. Furthermore, they will make it possible for us to focus on more value-adding tasks. And happy employees lead to happier customers.

Here's the point: Handling routine calls, data entry, report preparation, log keeping, and documentation is a waste of time for humans. These tasks downplay human interaction, the most fulfilling part of most jobs. For example, once workers in an insurance contact center realized that the new virtual assistant wouldn't take their jobs or increase their call-handling time, they found that getting advice from a chatbot gave them confidence, made the calls more interesting, and allowed them to address harder or more complicated exceptions.

The metaphor of the head up front and the machine in back makes sense to most people. Humans and machines have clear advantages at the extremes. Machines will always be better at transactions, and humans will be better at judgement, leadership, and conversation. But we need significant upskilling to address the middle of the centaur. Unfortunately, that's the area we know the least about. The World Economic Forum surveyed the business community in 2018 and found that no fewer than 54% of all employees will require significant reskilling and upskilling, with most of it needed for new human-machine collaboration.[116]

Humans And Machines Will Become Harder To Distinguish

Let's all agree that we hate robocalls. They get us on our mobile devices. They call us at work, at home, and everywhere in between. Soon we'll have bots that will fight off robocalls on our behalf and thwart the soulless companies that deploy them. The point here is that robocalls get more human every day. They may start with a giggle and a flirty quip about adjusting a headset. Machines are evolving to be more human. Royal Caribbean Cruise Lines worked with Marco Pelle, a principal dancer at the New York Theater Ballet, to design robot bartenders to take on human traits.[117] They're short on relationship advice but use 30 spirits and 21 mixers to produce 1,000 drinks each day. Ice cream robots on cruise ships may be next.[118]

Even today, experts in software robotics recommend we that govern digital workers, software bots that perform work for humans, in the same fashion as human workers. This means tracking their hire date (software creation date) and assigning a boss (responsible for design, training, and securing the bot's password access). Each bot will even have a performance review and a termination date (when they're taken out of service). In this way, governance and management of digital and human workers are converging.

Want an example? NASA has named its digital workers after former presidents to make them part of the team.[119] Washington is a full-time digital employee, a 24x365 software bot built with an RPA platform. He has a login and password, a unique identifying number that NASA can audit, an email address to send and receive work, and a supervisor to assign work instructions. He can learn new instructions in 1 minute and be trained in 2 to 4 weeks to perform any task that uses a keyboard and mouse.

Machines will take on human characteristics, but in addition, humans will become more machine-like. Take employees on the phone with a customer.

They can now search, correlate, and, with AI, gain insights into a customer's behavior. The machine makes them appear far smarter then they deserve to be, almost like giving them a virtual Superman's cape. And what about customers? They now take on machine characteristics as well. They have access to social media, the World Wide Web, and specialized apps that give them unprecedented power and insight.

We don't even know what the centaur's middle will look like in a few years. The blending of humans with machines will inevitably treat the worker in a less human, more metrics-obsessed way. Humans interacting with machines requires AI ethics, knowledge curation, explaining, bias assessment, and human protocols that are still in the early stages of development.

Training, for example, now depends on mentoring, which requires a human being who can understand our strengths and weaknesses. When we lift the human out of a process, we lose this unique training ability. In medical training, an attending doctor does the more involved procedure while a second attending intern or resident observes. Each medical professional judges the progress of the next in line. But how well does this process work with an image-guided surgical robot in the middle? Should a human or a machine mentor a robot? Or should both?

Stress Monitoring And Management Will Become An Important Middle Skill

One question that's largely overlooked today is the effect of automation on employee attitudes, emotions, and anxiety. The American Institute of Stress says that stress is already the No. 1 US health problem and further states that job-related issues are the No. 1 cause of stress in adults.[120] Toxic workplaces may already be responsible for 120,000 deaths per year — which would make workplaces the fifth leading cause of death.[121]

AI will create more stress for employees if employers don't manage it proactively. As workers see automation eliminating positions, they become anxious and protective of their jobs. If the government provides less of a safety net, they will nervously look to the talent economy. We need to understand the psychology of workers who now take direction from machines rather than people. Here's David Johnson, principal analyst at Forrester and a top authority on employee experience:

> No question, automation will lead to more monitoring
>
> of worker behavior. They'll take in voice, video, and
>
> image feeds. They will set schedules, compare a worker's
>
> output with others, and will be quick to determine your
>
> status. But this can be positive. Honest workers, for

example, are happy to have proof they're not stealing food at the restaurant.

But most of all, people are happiest at work if they are making progress.[122] If an automation helps them do the job better, it's all good. But if the automation doesn't, then anxiety results, especially if they have lost control and insight into the process due to expansion of the machine's role. Think of this way. A bakery starts out making bread from scratch. With expansion (scale) it starts using a mix for many core ingredients. It's not long before employees forget the original ingredients. The nuances of a process, now scripted in an algorithm, soon become lost to the human workers. They work around the edges, have full accountability, but no longer control or have insight into critical steps.

At the end of the day, it comes down to trust. If they trust the machine to make the right decision, the experience will be positive. If we show them the right data that supports a machine decision, they will trust the machine and have less stress.

Here's an important takeaway: Today, there's no measurement, reporting, or requirement to consider stress for workers. Enterprises mostly focus on performance metrics that leave them blissfully unaware of what they're doing to employees. And right now, employees are trending anxious. In 2017, the Pew Research Center reported that 72% of Americans are worried about the impact of automation on jobs.[123] Incredible as it may seem, this survey was taken with unemployment rate at a 20-year low.

9 THE TALENT ECONOMY

The gig economy exploded on the scene to support the sharing economy but is evolving to something more, a permanent and growing addition to the traditional workforce. Millions of workers across the 12 personas will move to it. And many will be bargains. Companies in developed economies, like the US, pay a 25% to 40% premium for the direct employee versus the gig worker.[124] Government-mandated taxes, vacations, healthcare, and other benefits drive this difference, and it's likely to continue. But that's just the economics. The bigger change will be cultural. Contract work, in all forms, will become more accepted. Yet, despite these trends, most businesses still operate as if the classic model of owning talent will dominate. It won't.

The Current Gig Economy Leaves Much To Be Desired

Traditional employment is a constant validation of one's worth. It augments home and social life in important ways. The 9-to-5 workday, employee badge, business card, regular paycheck, and pecking order at work, while perhaps painful, provide a stable foundation. Rooting for success, company events, the softball team, March Madness bracketology, and workplace friendships all reinforce the extended family.

The current gig economy falls short in these and other respects, according to Rich Lane, who has lived in the two worlds of work. He worked at Splunk, a provider of machine learning and analytics software that predicts problems in advanced computer systems. Prior to his time at Splunk, Rich spent several years in various technical support and management roles and traveled the world in the US Marine Corps as an avionics specialist. Here's his story:

> My dad was a Korean vet and worked in the same factory
>
> for over 30 years. When I was a kid, he would take my
>
> brother and me there and we thought it was just so cool,
>
> big, and important. All that huge machinery and people

running around, but he told me, "This place is dirty. Its noisy and smelly. You're eating dust and dirt for 8 to 10 hours a day. You need to do something better." I was 9 years old and had no idea what he was talking about.

Later, he took sick. My brothers couldn't really help. They had kids and worked hourly jobs but I had no kids, so it came down to me. And that's when I entered the gig economy. I worked for Amazon driving around packages — all shift work and luck of the draw. No one gives you work; you must go get it — like an Uber driver grabs a fare. And you must hustle, get there first thing. It's hard for some. I was good. I could get 4 hours of delivery credit in 2, and that raised my hourly rate. But others couldn't figure it out as well.

They are a data-driven company and scale is everything. There is very little human presence. An algorithm is your boss. If you do one thing wrong, some metric will catch you, and you're just gone. It's not like they say, hey you're a good driver, we'll give you another chance. There's zero loyalty and no appeals process. They have so many people trying to get the job, they don't need one. They need to keep the machine going with as much unskilled labor as they can get.

We glamorize the gig economy. It's painted as freedom, good money, and being your own boss. But it only works for people with a modest and ratcheted-down existence. Maybe they use a partner's healthcare plan. I'm a lucky guy. I've got a computer background. And I saw how the gig economy didn't work there either. When you outsource technical work, the person or team you hire has little vested interest. If I'm a Unix admin, I may have 50 different clients to serve, and I can't possibly care much about each one. You just want to fix the issue and move on. But as an employee, that server in that data center is your baby. I'm going to worry about it more, talk to the storage guys and see how I can prevent problems.

The gig economy isn't great, but neither are a lot of companies. They need to be more introspective. They should say, yeah, we're an insurance company. It's not cool and sexy, but we can make it so people don't hate to come and work here. A few of my friends have just given up. They just want a job. I'd say, you know you'll never get anyplace there. And they'd say, that's OK. I like leaving here at 5. Go home, play with the kids, have a beer, watch a ballgame. I'm good with that. That's how I want to lead my life.

Here's the point: The gig economy has been developing for decades. It's far from new. Contractors, part- timers, and consultants have toiled for some time. The total number of gig workers in the US is already at 57 million, or 36% of the total civilian labor force.[125] Ninety percent of employers already use independent contractors to gain access to workers with specific skills as the need arises.[126] But the current gig economy won't work for the large number of automation deficits that will enter it. We need something different: a talent economy.

We Must Prepare For Millions Of Fresh New Entrants

Digital elites will do the most to reshape the gig economy into something else. They have the most influence. For them, the word "job" is less relevant than clients, contracts, projects, and tasks are. Millions of Baby Boomers will also enter the talent economy. Ten thousand are retiring each day, the population of a small city closing the door on decades of experience.[127] It's called the "silver tsunami," and we're not prepared for it. Automation hasn't yet protected a firm from the loss of institutional memory and potential intellectual property as employees retire. Do we still need COBOL programmers? Yes. Lots. Corporate amnesia is now a category in some risk-management frameworks. This opens the door for many Boomers to pursue contract work to supplement fixed incomes and enjoy social interactions outside the home, but only on their own terms. Before addressing his tee shot at the golf club, one retiree quipped, "I still work 7x24. I shoot for seven months a year and try to work at least 24 hours a week." Here's the bigger question: Is retirement a concept that should itself retire? With longer life and more opportunity for work-life harmony, is retirement necessary in the 21st century?

Even Art Can Make Money In The Talent Economy

Artists fit into the cross-domain knowledge-worker persona and are a perfect fit for the talent economy. Pandr Design Co. is a boutique studio based in Southern California. It builds custom art pieces that boost the social media presence of a business — huge murals, painted sides of buildings, and large amazing art displays. Its success represents the art of the possible in the talent economy. The winning formula: two women, lots of talent, controllable fear, and Instagram. California native and burrito lover Roxy Prima, with her colleague Phoebe Cornog, is the cofounder of Pandr Design Co. Here's Roxy's story:

> I'm a trained graphic designer. The commute to the graphics company was terrible. The work boring. And then there was the glass ceiling I kept hitting my head against. It was a small firm but with many stale stereotypes. I kept quiet about these things. I know I had more talent and drive than they could deal with. Then one day it all boiled over. I ditched my office job that had me locked behind a computer screen all day. I traded it for a pencil and a big paint brush. And life as an entrepreneur.

I wanted to prove just two things: One, you could make money doing what you loved. In this case art, that we could combat the "starving artist" stereotype. Two, that women could build a successful business. Sometimes we go into Home Depot to buy gallons of paint. The contractors don't know what to make of us. They don't know what to call us. But Instagram was the key. Once I got to a thousand followers, the ability to network just grew. We could post our best work, top clients, and got traction quickly. And best of all, we don't work with clients we don't like. We never tried the more specialized task-matching platforms for graphic artists just because Instagram worked so well for us. No question, there is just a lot less stigma to going out on your own now. It is much more acceptable to not have a traditional job. Even my parents are now believers.

There are several points to the Pandr Design Co. journey. For these two, work moves from stint to stint, creating bursts of exciting moments that they can field through social media. The focus is on marketing their personal brand and talent. They've replaced loyalty to a company with passion for what they do. The talent economy also fosters a different view of success. Employees in traditional jobs come under the growth obsession, the economy's ingrained measure of success. Employers manage them to metrics that increase revenue, profit, or the number of items they produce. In the talent economy, the best measure of success isn't always growth. For Roxy and Phoebe, producing consistent quality is as important as growing big. Bigger isn't always better. Higher quality, brilliant design, and delighted customers are. Just being better is — better.

Cubicle Workers Will Take To The Talent Economy

"Thanks for holding. Your call is very important to us." But apparently, it's not all that important. If it is, why isn't anyone answering your call? As we showed in chapter 6, software robots will eliminate many of these jobs, and many will move to the talent economy. Today, most contact center employees are full-time. They have a high-school diploma or equivalent. They have benefits, report to a fixed location, and make $15.81 an hour.[128] Chatbots that take on customers directly will reduce full-time positions; software robotics will handle repetitive tasks. But many people will just move to the talent economy, and why not? It lets workers stay at home, lowers employer costs, and reduces carbon emissions.

Those workers might also do a better job. The talent worker is often compensated on results (e.g., they're paid only when they solve a customer problem). Traditional employees are paid whether they've helped a customer or not. The high-performing talent worker may solve more problems at a greater rate and earn more than full-time rivals. Being older and better educated helps as well. Virtual customer support workers are, on average, 38 years old versus 23, and they have more schooling.[129] Talent workers are often hired for their product experience. Customers trust real users of the product. Maybe the company recruited the worker from one of the many online communities that help solve problems for free.

Attitudes Shift As The Talent Economy Expands

Full-time employment and career paths have defined many of us. Traditionally, a good career raised your marriage prospects and gave your parents and future in-laws something to talk about. A casual view toward full-time work might lead to discomfort, a loss of eye contact, and qualifications like "my boy is still finding himself" or "my daughter is carefully exploring options." If you described yourself as a consultant, friends would politely assume you were unemployed and looking, refraining from further questioning.

Hard-working families are idealized. Announcers laud the athlete who works hard off the field. Digital elites show off their epic work schedules. Even severely disabled welfare claimants must seek work; 2017 federal policy changes now require that those receiving Medicaid, even the very sick, meet work requirements to receive health benefits.[130]

These attitudes about work develop at an early age. Raymond fits into the coordinator persona and lost his job of 10 years at a Midwest bank:

> Work helped me grow up. It defined my transition from
>
> a kid to an adult. At some point you had to become
>
> responsible. Show up here or there, on time, deliver
>
> something. Suddenly you had to have your act together.
>
> You had an alarm clock to respond to or tools to put

away or a shop to close at night. It made you feel more human.

And this started early. Helping your mom in the kitchen or your dad fix the car. The threat of a poor job was my parents' ultimate weapon. My dad would say, "If you don't study in school, you'll be a trash collector." It was your life and your work and with it, you defined yourself. You made something of yourself through work. And this meant a job, a profession, a regular paycheck.

But attitudes like those of Raymond's parents are changing. Acceptance of people who don't have traditional jobs is on the rise. Bob is an example of someone whose attitude about work has slowly changed:

I used to have this great fear, that by leaving work I would become less of a person. This was the doctrine that kept me working for so many years. Ozzie and Harriet, for those old enough to remember, set the tone. The clean white collar, home at 5 every night. That was part of being head of the family — contributing something. I think my favorite part of the day was driving to work, seeing all the commuter traffic, being part of the great rhythm of work. And you get the picture. But now a new social message is creeping out. You will not be disowned by society if you do not work in a regular job. I really left work twice. The first time was when silently in my head I became open to the possibility, when I decided my life was incompatible with my work. And a second time when I physically resigned to take a contract position.

Here's what is no longer true. Keep your head down, work hard at a company, develop valuable expertise, and you will be rewarded based on your merits. This promise of stable rewards due to tenure are no longer true. There's been too much downsizing, Wall Street-driven mergers, and acquisition. Loyalty from the company to employee, and the reverse, no longer exists. Traditional career paths are just not reliable.

Bob represents a shift in workers' attitudes. He depends less on companies. A personal reputation will provide more income security than a resume with five-year stints. You're better off replacing your top-down and bottom-up company loyalty with your reputation among peers, clients, colleagues, former mentors, and teachers. These networked resources are often deeper, more stable, and more committed than your managers and coworkers, all of whom may be victims of tomorrow's merger. Airport bookshelves are littered with books on how to make a name for yourself and create a personal brand.

Talent Access Will Surpass Talent Acquisition

Digital elites with skills in high-demand areas — like data science or digital marketing — are choosing to work for themselves, and enterprises will have no choice but to accommodate them. Walls between internal and external talent will soften, and emerging platforms that match skills to tasks will be the softening agents. Systems to onboard new employees, administer benefits, manage performance, and acquire talent, all built for the old view of work, will decline in importance. How relevant are these when half a company's workforce will be from the talent economy? The semantics for traditional job descriptions, personal actions, resumes, reporting structures, and organization charts will soon be legacy artifacts.

Ironically, automation, after disrupting current work life, will be the glue to formalize the talent economy. First-generation freelance platforms like Fiverr and Upwork are adding new platforms and will advance to match skills to tasks for writers, data analysts, proofreaders, and researchers. Already, you can find a lawyer on Fiverr to do a trademark search for $10. You won't know details about who or where they are, but they won't charge hundreds of dollars an hour. Patients Beyond Borders matches patients with lower-cost healthcare options. The FOA will create automation deficits, but automation will also formalize the talent economy.

What About The Informal Gig Economy?

There are two parts to the gig economy, one that people talk about frequently and one that they largely ignore. The formal segment is what we think about first. It has some form of contract for employment and will evolve into the talent economy. But far more people are in the informal gig economy, and more will follow.

So, what is this informal economy? It includes street vendors in Mexico City, rickshaw pullers in Calcutta, pushcart vendors in New York, garbage collectors in Bogotá, and roadside barbers in Durban. They're less visible workers who repair motorcycles, recycle scrap metal, make furniture and metal parts, and work in restaurants and hotels. They're subcontracted janitors and security guards and casual day laborers in construction and agriculture.

They aren't monitored by any form of government, and they may not pay taxes. Unlike the talent economy, activities of the informal economy aren't included in the gross national product (GNP) and gross domestic product (GDP) of a country. There will be a steady climb in these numbers. The African Development Bank Group estimates that sub-Saharan Africa's informal economy accounts for 80% of its labor force.[131]

10 IT'S DIFFERENT THIS TIME — IT REALLY IS

As described in chapter 1, you can reduce future-of-work beliefs into three views. The historians believe that AI will be absorbed into normal work patterns, as previous automations have been. It will be business as usual. Workers will shift to different jobs without changing the total number employed. Robot enthusiasts take a second view. They see workers being freed from mind-numbing tasks that are beneath humans, anyway. And then there are the dystopians, who believe that robots will take all our jobs and head to the streets.

These views are black or white, but the future isn't. We'll see a long period of gradual disruption to the traditional economy that anchored society. It will be a slow, rising tide of automation that affects every job and worker. Cubicle workers and coordinators are the early casualties, but waves will occasionally splash up to higher levels and dampen previously secure knowledge workers (see Figure 10-1).

So, what do we do about it? To answer this question, we spoke with more than a hundred different workers across industries and with countless tech industry executives, businesses, academics, and government organizations. We combined these findings with Forrester's take on AI's progression and future-of-work framework and model. Putting all this together, we concluded that this age of automation is different, and we need to deal with the difference. Here's a summary of the top reasons why and some of the best ideas to deal with it.

Why It's Different No. 1: We Have These Things Called Software Robots

When people think of robots, they imagine a human form. Or they think of a machine designed to perform a task, like a driverless delivery vehicle or a machine racing around a warehouse. Automations, in most people's minds, are physical. Software robots are different. They're invisible and more agile. They can learn faster and benefit more from accumulated knowledge than their hardware brethren. They're a force we haven't reckoned with. Software robots are critical automations in all but the physical worker (288 occupations) and human-touch worker (76) personas. Indeed, the robots that will take the most jobs in the next 10 years will be those you can't see. And here's a near certainty: By 2025, most workers will have a personal robot, one they can configure by themselves through text, using emerging forms of self-configuring machine learning.

Figure 10-1 Water Will Splash To Higher Levels As The Automation Gains Intelligence

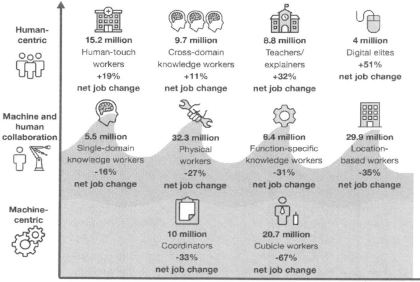

The water rises as automation handles greater context and variability

Note: Numbers of workers are totals for today's workforce. The percent automated includes the deficits or jobs removed from the workplace by 2030. Automation dividends will offset these percentages.
Source: US Census Bureau annual projections for the US up to the year 2030 and US Congressional Budget Office labor-force participation-rate projections for the US up to the year 2030; automation percentages are Forrester forecasts.

What Can We Do: Identify And Promote Constructive Ambition

The most disruptive form of AI will be software robots. Some will combine robotic process automation and different AI components. Others will simply be algorithms running in machines that make decisions that humans used to make. Still others will be chatbots that help solve customer problems. These forms of automation will create deficits, particularly in three work personas: cubicle workers, coordinators, and function-specific knowledge workers. We need to help these people transition to better jobs. We need to encourage constructive ambition, which is the quality, based on an inner desire to achieve and a belief in oneself, that helps us reach success. It exists in all of us. We need to get better at finding it in individuals and turn potential deficits into dividends.

Why It's Different No. 2: Previous Automations Left Humans In Control

Previous automations largely left humans in charge. Carpenters benefited from the use of nail guns, but they still decided where to put the nails. But greater quantities of data, processing speed, and machine learning algorithms will push decisions from humans to machines. We've documented the unprecedented control issues that result: Humans have less control and lower self-esteem, management cedes business control to the "black box," people give up control of their personal data, and workers lose control to a new class of digital elites.

What We Can Do: Develop Our Robotics Quotients

Business needs to develop risk-management processes to control unpredictable events created by algorithms that make decisions based on probability. Bias and reliability in training data, along with the control of data to ensure privacy, must become a priority. J.P. Gownder developed the robotics quotient (RQ) framework at Forrester:

> Most companies don't have the competencies to implement automation technologies successfully. To thrive with automation, they need to develop and measure their robotics quotient. The RQ is an assessment framework that measures the ability of individuals and organizations to adapt to, collaborate with, and drive business results from AI, automation, and robotics. The RQ model helps leaders self-assess their fundamental capabilities, across their people, leadership, organization, and trust. As automation, AI, robots, and other intelligent machines change how employees conduct their day-to-day work, companies

must take on the challenges of human/robot collaboration. They must also determine trust: The trust component of RQ evaluates the opaqueness, governance/auditability, and human effects that automation will engender. RQ will help solve new problems, like humans working with probabilistic machine-generated insights and decisions.

Why It's Different No. 3: Automation Will Creep Into Knowledge Work

Previous automations largely focused on the cubicle and coordinator personas. But AI is promiscuous. It affects all 12 personas to some degree. Cross-domain knowledge workers like trial lawyers and economists will be less affected. Machine algorithms can't make the types of decisions they make. AI's inability to make connections with variables outside the defined data model will prevent it from disrupting the upper echelons of knowledge workers.

But lower-echelon knowledge workers, like single-domain knowledge workers, are more vulnerable. For the first time, they're not safe from becoming deficits, and many will go through significant job transformation. Function-specific knowledge workers such as insurance underwriters are in real trouble. All knowledge-worker personas will lose well-paid positions that support them with tasks like gathering and organizing data. These jobs will move quickly to machines.

What We Can Do: Prepare To Exploit The Machine's Weaknesses

Recognize that machines have limitations, and prepare to show the machines who's boss. They can process far more data at greater speeds, but they can't select which data or variables are most important. This is a huge limitation, and it's where both single-domain and multidomain knowledge workers will provide the most value going forward. Those preparing for a career should focus less on mastering a body of knowledge that will become part of a machine knowledge base and more on how to construct the decision framework that machines will use.

Why It's Different No. 4: Mission-Based Workers Become A Thing

Trust in traditional institutions like government, religion, and community is breaking down. Employees will look to companies to pick up the mantle of social responsibility. Many workers will adopt a mission-centered existence, where what people believe in and care about is more important than their net worth or material possessions. Prospective employees will choose their next employer based, in large part, on the brand's commitment to corporate social responsibility as the job seeker defines it. In 2008, Peggy Noonan summed it up well in a *Wall Street Journal* article on the life and legacy of Tim Russert, titled "A Life's Lesson":[132]

The world admires, and wants to hold on to, and not lose, goodness. It admires virtue. At the end it gives its greatest tributes to generosity, honesty, courage, mercy, talents well used, talents that brought into the world, make it better. That's what it really admires. That's what we talk about in eulogies, because that's what's important. We don't say. "The thing about Joe was he was rich!" We say, if we can, "The thing about Joe was he took care of people."

What We Can Do: Enterprises Can Take A Stand

Mission-based workers will aspire to more than just regular paychecks. Enterprises need to get more comfortable speaking out on real issues. Most aren't. Companies need to take sides in the culture wars — think Nike and Patagonia on one side and Chick-fil-A on the other. Expect to see a few brands trying to differentiate themselves by following the lead of Nike's successful 2018 ad campaign.[133] Corporations should develop a strong moral center. The growth in the mission-based persona will make it pay off for them.

Why It's Different No. 5: The FOA Will Restructure Education

Previous automations didn't challenge the educational system the way the FOA do. Workers now need education of a different kind. The rapid pace of automation and our growing life spans will soon make our current approach obsolete. Children born today will likely live to be 100. Today's life pattern of learn, work, and retire will be replaced by learn, work, break, learn, work, break, and it will extend for decades.

What We Can Do: Encourage The Private Sector To Step Up Its Education Game

The private sector needs to take a bigger role in education. Traditional education will struggle to deal with automation's refresh cycle and will have a declining value proposition. Even today, universities are seeing a decline in college submissions.[134] Dr. James Tracy studies the future of work at the Woodrow Wilson Institute and told us that traditional college education is being oversold. We don't need four years of college to accumulate deep expertise in multiple areas:

> Knowledge will be managed by machines. It is less of a
>
> competitive advantage. We need to spend less time
>
> mastering content, but just enough to make those cross-
>
> domain connections. Teamwork, empathy, value,
>
> judgement, leadership needs to be the focus.

Work experience with certifications must become more central. With 2.4 million employees, Walmart is the largest employer in the world and has invested 2.7 billion dollars with that in mind.[135] Two hundred Walmart academies are up and running, with some in old warehouse space that's empty due to supply-chain automation. They're using new techniques. They have 17,000 Oculus Go Virtual Reality headsets to develop customer service, digital, and compliance skills by, for example, simulating a difficult encounter with a customer. At the end of training, employees don caps and gowns and receive certificates on a stage. One woman said, "I never graduated from high school. This is my first opportunity to graduate from anything, and for my children to see me be successful at anything."[136] These programs teach workers how to be better Walmart employees. Fair enough. But an even broader curriculum would develop generic and transferable skills to help them to move into the middle class.

The way forward requires a different approach to training and education. The only way individuals can be robot- or automation-proof is to adopt a lifetime learning attitude. To support this attitude, the private sector and our educational system need to collaborate on a cradle-to-grave endeavor.

Why It's Different No. 6: Automation Anxiety Will Play A Big Role

Interviews confirm that automation and AI affect workers' attitudes in new and different ways. Attitude, emotions, and collective psychology will play a big role in the restructuring of work. Previous automations reduced toil and made work faster and easier. Electricity gave us light and energy, trains got us places faster, and the internet allowed us to locate and obtain things more easily. But AI will start to boss us around, expose skills gaps that result in digital insecurity, and challenge us to keep pace. And these are just the direct consequences of working with intelligent machines.

Indirect consequences from scale, convergence, and control progression dig deeper. For example, if we no longer trust an employer to return loyalty, we in turn give them none. If we have a heightened sense of material insecurity, we reduce retail purchases and save more. If we see automation rapidly eliminating positions, we become anxious and more protective of our jobs. If we don't trust our government to provide a safety net, we look to the talent economy and focus on a personal brand or portfolio of capabilities. Our beliefs about work will change. We've left the golden age of work in the dust, along with the belief system that supported it. The restructuring of the workforce comes with a recasting of our perceptions. And with something more to worry about — automation anxiety.

What We Can Do: Take Automation Anxiety Seriously

First, we need to acknowledge that automation is leading to economic stress and a new form of digital insecurity. Recognize that metrics obsession will degrade and demoralize many workers. The surge in available data can tempt employees and managers to focus more on the metrics and less on improving the employee experience. We need identification and treatment options that deal with machine and work stress issues.

Stress and anxiety at work is a major issue in Japan.[137] The Japanese even have a word for it, *karōshi*, which means "death by overwork." The major medical causes of karōshi deaths are heart attack and stroke due to stress and a starvation diet. In December 2017, Matsuri Takahashi, a 24-year-old graduate recruit, took her own life. She toiled at Dentsu, an advertising company. She logged, toward the end of her life, more than 105 hours of overtime a month, almost three extra weeks of work. The government found that one-fifth of Japanese companies admit that their staffs work dangerously long hours. We need to figure out how automation can ease rather than exacerbate the workload of our workforces and our supercharged productivity cultures.

Why It's Different No. 7: Inequality Fuels Global Unrest

Automation has never been so laser-focused on the middle class. The FOA will reduce middle-class bargaining power. Existing workers will end up in lower-wage jobs. Without rising wages, the dreams of families worldwide to live in good homes, support their families, retire comfortably, and see their children do better won't come true.

Wage stagnation adds to automation anxiety. Workers feel discouraged and become less likely to explore new economic activity: starting a new fast-growing business, switching jobs, or moving across the country. More and more, we'll see them take to the streets. The recent "yellow vest" movement in France is a cry for help from a distressed, frustrated, and furious middle class.[138] What we're beginning to understand is that income gaps and wage stagnation are as big an issue as job loss.

What We Can Do: Stimulate Wage Growth And Give Human-Centric Workers Their Due

Minimum-wage legislation sets a floor but does nothing to promote wage growth. We need to add policies that target wage growth and reverse those that discourage investment in people. For example, our tax system favors capital investment (depreciation allowances) but doesn't reward investment in labor. This encourages businesses to invest in that new machine or new AI system but not in people.

But here's something we can do without political discourse: give workers in the human-centric personas the respect they deserve. Empathy and communication skills are their core talents, but they're undervalued. Enterprises need to treat workers in these personas better. It's not only the right thing to do — there's an economic argument in favor of providing better training and jobs for employees across all personas. Evidence is mounting that low pay, few benefits, unpredictable schedules, and dead-end careers lead to bad employee attitudes. These, in turn, degrade the customer experience, hurt business, and — in extreme cases — lead to civil discontent and political unrest. We can avoid these outcomes.

Why It's Different No. 8: Location-Based And Physical-Worker Personas Need Love

Drive through any suburban area and stop at any modest cluster of shops and offices. Look at the roadside signs. You'll probably see small offices for plumbing and HVAC contractors, an electrician, a small bakery, and a dry cleaner. These commercial spots employ millions and will take on characteristics of the talent economy, allowing the building of a personal brand that's spread through word of mouth and social media like Angie's List. Fred Goff is CEO of Jobcase, a social platform for the future of work. Jobcase was formed to empower the world's workers in their work lives. The approach centers around community, not just networking. His view is that regular workers can prosper:

> We have 100 million workers on Jobcase. They are hardworking, veterans, and patriots, the heart and soul of working America. And they need help. They are anxious and concerned. They have seen decades of stagnant wage growth with automation advancing rapidly. Only 25% think K-12 will prepare them for the future of work. They believe employers will have knee-

jerk layoffs and move full-time staff to on-demand labor. Our people need to watch their backs. I have 25-year-old workers on my network that have had five jobs already.

Self-esteem and positive attitude development are our most important goals. Look, I had two hard-working parents. They told me over and over I could do anything or be anybody I want. This is not the case for most. They need empathy, a simple message from another human, you can do it. I hear people say, there is no hope for the working class. Robots will take their jobs. High-school graduates aren't going to move to the digital economy anytime soon. I tell them that's condescending. They are snobs. They can be a Salesforce administrator and make $50K a year. 90% of our people can do this. It's not all about starting to code to lift income. Longer term, it's about solving problems, fixing things and that's what our people can do.

What We Can Do: Encourage A Wider Scope Of Acceptable Jobs

The number of human-touch workers like health aides is growing, and most are women. But today, men who lose a factory job don't rush to become health aides. Many see it as women's work, which has always been devalued in the American labor market. This attitude must change. And robots will start doing the heavy lifting and will open traditionally male jobs to women, as with Nissa Scott at the Amazon plant. Let's dismiss age-old job stereotypes.

As a society, we need to stop looking down on workers in the physical, location-specific, and human-touch personas and looking up to cross-domain knowledge workers in law and finance who provide less tangible value. Not everyone needs to, or should, go into six-figure debt for a college education that machines will marginalize. So why do we discourage our young from going into small-business trades? Why do guidance counselors and parents disparage working with your hands? We need electricians and plumbers, and lots of people can be happy doing these jobs. A robot to diagnose and repair a plumbing problem in an older home will be far too expensive. Clean-energy jobs like solar panel installers now number more than 350,000 and will grow at double digits.[139] We don't have enough people to do them.

Why It's Different No. 9: Previous Automations Didn't Restructure Private Sector Hiring

Previous automations didn't structurally change how companies recruited, retained, and managed resources. Employers shifted workers to new jobs but left the basic structure of work intact. A telegraph technician turned into a phone operator but still worked for a company, used transferable skills, and didn't see dramatic changes to wages. Today's automation alters the very structure of work. The workforce will be a mix of full-time employees and outside talent with no formal ties to a company. Workers will move from role to role and across organizational boundaries more freely than ever. Enterprises, cultures, and systems today are locked into an "owned talent" mentality and aren't prepared for this shift.

What We Can Do: Leave The "Owned Talent" Culture Behind

Enterprises will find that their once-stable workforces will begin to contract, expand, and shift as the talent economy matures. Businesses will move from a structure where humans are hired and tiered vertically and migrate to one where the talent economy and r robots do much of the work (see Figure 10-2).

Corporate America will have to adjust to a new way of recruiting and managing people. Enterprises must convince employees that working for that company, wherever it may take them, will be good for their careers. And the only way to convince them is to support the continued learning and development that an automated workplace needs.

Enterprises can broaden their view of automation. They must invest in AI not only to reduce costs but also to lift employees to new levels of commitment, energy, and productivity; provide a more human face to a brand; and take customers to new experiences. An automation framework can help. Glenn O'Donnell is a research director at Forrester and spearheaded our recently published framework:[140]

> Many of the issues that AI introduces and that we
>
> discuss in this book can be better understood with an
>
> automation framework. And we built one. It has nine

dimensions that allow you to profile any planned automation for its effect on process, enterprise, and people. You can understand how much learning the automation has and how transparent or opaque it is to deal with black-box scenarios. You can rate the "people" effect to see how your workers are affected and figure out how you're organized. We need enterprises to take a broader view of automation, and they need a modern and structured approach to do so.

Figure 10-2 A Burstable Workforce Supported By Automation Is The Future

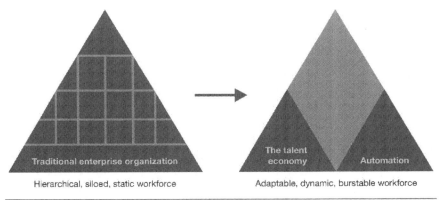

Hierarchical, siloed, static workforce   Adaptable, dynamic, burstable workforce

Why It's Different No. 10: Government Has Never Been More Out Of Touch

Government lacks a realistic and practical view of automation from the worker's perspective and an actionable framework to drive policy. Their estimates of job growth don't account for automation trends or the emergence of the talent economy. They still care more about counting farm and non-farm workers — an antiquated perspective. For example, they overstate projections for security officers, fleet managers, and other logistics occupations, and that sends the wrong message to those entering the workforce.

What We Can Do: Target High Automation-Deficit Areas

Two areas of direct support make sense today: targeting areas with huge forecasted deficits and creating a market-based robot or automation "reserve" to support the transition.

Based on our research, we can be sure of this: The number of digital elites will grow in major urban centers like Austin, Boston, New York, and San Francisco. Appalachia, Midwestern and Rust Belt states, and areas lacking a strong university system will experience automation anxiety, a growing income gap, and resentment. We need to launch a "digital Keynesian" movement.[141] Instead of propping up the economy with defense spending, let's advance digital investment to subsidize job development in disadvantaged areas.

Here's another idea: Let's be honest about automation deficits. Economists might call these "negative externalities," a concept developed in the early 20th century by the British economist A.C. Pigou.[142] He noticed that trains emitted embers from their stacks that set farmers' fields ablaze. He recommended that the railroads pay into an escrow fund that would compensate the farmers. This made the railroads aware, in a tangible way, of the implications of their actions. In a similar way, companies would pause and think through the people effect of their automation ambitions.

Here's how it might work: Firms that replace humans with robots (software or hardware) would contribute to an internal fund, in an amount based on their annual automation deficits. Our future-of-work model can help calculate these amounts. Automation dividends — the new digital jobs created — would be subtracted from the deficits. Net deficits would then be used as a basis to set aside funds to support lifetime learning or provide wage insurance. But this wouldn't be a tax going to the government and back to citizens through programs, and it wouldn't be a politically charged "guaranteed minimum income." Rather, the private sector would keep the dollars but would have to follow government guidelines to use them. Bill Gates is a supporter of a robot tax plan.[143] He agrees it may slow "progress" down a bit. But that's OK. We need time to think through the implications of control, scale, and convergence and how they affect people.

Automation For The People Requires Help From All

A hundred years ago, a factory could spew carbon in the air and dump waste in the river. No one said anything. Decades later, we realized it was bad for society and changed our ways. Let's not treat AI the same way. We can get ahead of it. We shouldn't look back 20 years from now and say, "You know what, we should have thought about how this technology affects people." We need to ask the hard questions — now. If companies need fewer workers due to automation, what happens to those who once held those jobs and don't have the skills for new jobs, i.e., the digital outcasts? And because jobs deliver many social benefits, how are people outside the workforce going to earn a living and get these healthcare and social benefits?

If we manage the wave of automation proactively, the result will be good jobs and improved life quality for many. After all, nothing is cast in stone. We get to select the data and the algorithms; we get to develop, train, and fine-tune the models. We can decide how much to rely on the AI system and when to apply humans. We can design better approaches to support workers with lifetime learning. Unlike nuclear physicists in the 1950s, we can decide how to use the power of this new automation. We shouldn't just blindly automate, automate, automate. There will be no Adam Smith "invisible hand" to sort it all out.[144] The time is now.

ACKNOWLEDGEMENTS

This was a difficult book to create, and I needed a lot of help. The future of work is a broad and complex topic — and an intimidating one. Its importance has led to many prominent, thoughtful, and diverse views. The effort has benefited from a collaborative culture, diverse perspectives, unselfish contributions, and colleagues willing to challenge as well as help develop the book's ideas and frameworks. And this describes Forrester to a T. I leaned especially hard on J.P. Gownder, a key collaborator on the book whose future-of-work model is at the very core of it. And without Glenn O'Donnell, this effort would have stalled very early. His enthusiasm and passion for all things automated was a driving force for the book project.

Forrester has a huge amount of knowledge on AI and automation, and I tapped into a lot of it. The early chapters on the forces of automation drew on research and insight from across Forrester, particularly from David Johnson, Martha Bennett, Chris Gardner, Ted Schadler, Andrew Bartels, and Frank Gillett. An important ingredient in this mix was Amanda Lipson; as the lead researcher, she masterfully controlled the chaos and kept things moving. And thanks to Diane Lynch, perhaps the best copyeditor in the business.

I'm also grateful for the full-throttle support of Forrester's senior leaders, particularly Carrie Johnson, Laura Koetzle, and, of course, George F. Colony. And finally, I'd like to thank my wife, Zaurie, for years of listening to me on this subject and for providing periodic blasts of much-needed social conscience.

ABOUT THE AUTHOR

Craig Le Clair is a VP and principal analyst at Forrester, where he has worked for over 10 years. He is recognized worldwide as a leading industry analyst on AI, robotic process automation, and the future of work and has authored hundreds of reports on the future of technology. In 2017, he received a Forrester research award for his definition of the software robotics process automation market, which was the top-read Forrester report of the year.

Prior to joining Forrester, Craig held senior positions at ADP, MITRE, and BBN Technologies. A prolific writer and speaker, he authored *How To Succeed in the Enterprise Software Market* in 2005. He is frequently quoted in *The Wall Street Journal*, *USA Today*, *Forbes*, and many other publications and media outlets. Craig earned a BS in economics from Georgetown University and an MBA from George Washington University.

METHODOLOGY

Most of the data and graphics in this book come from Forrester on the future of work. If you are a Forrester client with appropriate access, the cited address will take you to the report. If not, the Forrester website provides a brief summary of it. When citing a long web address, we use an equivalent address from the form http://forr.com/irqnX-Y. These site references were created for the convenience of the reader. Enter the web address into your browser, and you will be redirected to the appropriate site online. Please note that people sometimes change or remove content from web addresses that we have cited. The web content cited was visible at the time the book was written.

All occupational statistics are sourced directly from the US Bureau of Labor Statistics jobs database, which lists data for more than 800 occupations. We also reference the O*net database, which includes a breakdown of the skills required for each job.

Much of the information in the book comes from in-person, telephone, and email interviews by the author with workers and representatives of the companies described in the book. We made extensive use of human narratives from people of different ages, educational backgrounds, and demographics. As noted, some of these interviewees wanted to remain anonymous, and we have respected and noted that wish. Facts and quotes that don't have a cited source are from either public sources or these personal interviews.

ENDNOTES

Chapter 1

[1] Sapna Maheshwari, "Hey Alexa, What Can You Hear? And What Will You Do With It?" *The New York Times*, April 1, 2018, https://forr.com/irqn1_2. Consumer Watchdogs, a nonprofit advocacy group, examined patent applications from major cloud platforms, such as Google, and determined that parts of the application covered spyware and surveillance capability targeted at advertising.

[2] Shona Ghosh and Becky Peterson, "Buzzy AI software startup UiPath in talks to raise a mega funding round at a possible $7 billion valuation," Business Insider, March 19, 2019, https://forr.com/irqn1_3; Katie Roof and Rolfe Winkler, "SoftBank, Joining Robotics Race, Gives Automation Startup $2.6 Billion Valuation," The Wall Street Journal, April 29, 2019, https://forr.com/irqn1_3_2; and Jeremy Kahn, "Accenture Debuts Platform That Automated 40,000 Roles," Bloomberg, January 29, 2019, https://forr.com/irqn1_3_3.

[3] Time Chatterton and Georgia Newmarch, "The Future Is Already Here – It's Just Not Very Evenly Distributed," Interactions, April 2017 https://forr.com/irqn1_4. This a quote by American speculative fiction author William Gibson. It alludes primarily to the fact that the things that will constitute the normal, or everyday, within the lives of those living in the future already exist for some today.

[4] Craig Le Clair and J.P. Gownder, "Future Jobs: Plan Your Workforce for Automation Dividends and Deficits," Forrester report, April 30, 2019, https://forr.com/irqn1_5.

Chapter 2

[5] John Krafcik, "Where the next 10 million miles will take us," A Medium Corporation, October 10, 2018, , https://forr.com/irqn2_6.

[6] Adrienne LaFrance, "Self-Driving Cars Could Save 300,000 Lives Per Decade in America," *The Atlantic*, September 29, 2015, https://forr.com/irqn2_7. Researchers estimate that driverless cars could, by midcentury, reduce traffic fatalities by up to 90%. The primary reason is that most fatalities involve human error.

[7] Danielle Muoio, "Tesla's Autopilot system is partially to blame for a fatal crash, federal investigators say," *Business Insider*, September 12, 2017, https://forr.com/irqn2_8. The National Transportation Safety Board has said that aspects of Tesla's Autopilot played a role in a fatal crash involving Joshua Brown, age 40, in May 2016.

[8] Katherine Bindley, "The 22-Year Ride," *The New York Times,* February 27, 2009, https://forr.com/irqn2_9.

[9] "IBM Watson Jeopardy Mistake: What is Toronto?" YouTube Video, February 16, 2011, https://forr.com/irqn2_10. This short video documents the Watson error on *Jeopardy*.

[10] "What is Toronto?" IBM, , https://forr.com/iqrn2_11. Part of the explanation involved weighting: "Watson, in his training phase, learned that Jeopardy categories only weakly suggest the kind of answer that is expected, and, therefore, the machine downgraded their significance."

[11] Hannah Beech and Muktita Suhartono, "Confusion, Then Prayer, in Cockpit of Doomed Lion Air Jet," *The New York Times*, March 20, 2019, https://forr.com/irqn2_12.

[12] Russell Lewis, "Ethiopian Flight Data Shows Similarities To Indonesian Crash Of Same Boeing Model," National Public Radio, March 17, 2019, https://forr.com/irqn2_13. Ethiopia's Transport Minister said a preliminary review of the flight data reveals "clear similarities" in that accident and the crash of a Lion Air Boeing 737 MAX last October.

[13] "Operational Use of Flight Path Management Systems," FAA, September 5, 2013, https://forr.com/irqn2_14. This is a memo from the FAA with this finding. Jeff Wise, "The Boeing 737 Max and the Problems Autopilot Can't Solve," *The New York Times*, March 14, 2019, https://forr.com/irqn2_14_2.

[14] Johnson, "Pymetrics open-sources Audit AI, an algorithm bias detection tool," Venture Beat, May 31, 2018, https://forr.com/irqn2_15. This article reviews this open source "explainability" tool.

[15] Rory Cellan-Jones, "Stephen Hawking Warns Artificial Intelligence Could End Mankind," BBC News, December 2, 2014, https://forr.com/irqn2_16.

[16] Henry Blodget, "Mark Zuckerberg on Innovation," *Business Insider*, October 1, 2019, https://forr.com/irqn2_17. "Move fast and break things. Unless you are breaking stuff, you are not moving fast enough."

[17] Greg Sterling, "Data: Google monthly search volume dwarfs rivals because of mobile advantage," Search Engine Land, February 9, 2017, https://forr.com/irqn2_18.

[18] "The Top 20 Valuable Facebook Statistics — Updated April 2019," Zephoria, April 2019, https://forr.com/irqn2_19. Worldwide, there are more than 2.32 billion monthly active users (MAU) as of December 31, 2018.

[19] Larry Schwartz, "Wilt Battled 'Loser' Label," ESPN, https://forr.com/irqn2_20. Chamberlain, one of the greatest basketball players of all time, was often criticized for not winning enough, despite an outstanding statistical career. He often said, "No one roots for Goliath."

[20] Micah Singleton, "Nearly a quarter of US households own a smart speaker, according to Nielsen," The Verge, September 30, 2018, https://forr.com/irqn2_21. Nearly a quarter of US homes have smart speakers.

[21] Consumer Watchdogs, a nonprofit advocacy group, examined patent applications from the major cloud platforms, such as Google, and determined that parts of the application covered spyware and surveillance capability targeted at advertising. Sapna Maheshwari, "Hey Alexa, What Can You Hear? And What Will You Do With It?" *The New York Times*, April 1, 2018, https://forr.com/irqn2_22.

[22] Nicholas Thompson and Ian Bremmer, "The AI Cold War That Threatens Us All," *Wired*, October 23, 2018, https://forr.com/irqn2_23.

[23] Sam Shead, "China's president had 2 books about artificial intelligence on his shelf in his New Year speech," *Business Insider*, January 3, 2018, https://forr.com/irqn2_24. Pedro Domingos, *The Master Algorithm: How the Quest for the Ultimate Learning Machine Will Remake Our World* (Basic Books, 2011).

[24] Nicholas Confessore, "Cambridge Analytica and Facebook: The Scandal and the Fallout So Far," *The New York Times*, April 4, 2018, https://forr.com/irqn2_25.

[25] "Transcript of Mark Zuckerberg's Senate hearing," *The Washington Post*, April 10, 2018, https://forr.com/irqn2_26.

[26] Geoffrey A. Fowler, "No, Mark Zuckerberg we're not really in control of our data," *The Washington Post*, April 12, 2018, https://forr.com/irqn2_27.

[27] Tom Huddleston Jr., "Sean Parker Wonders What Facebook Is 'Doing To Our Children's Brains,'" *Fortune*, November 9, 2017, https://forr.com/irqn2_28.

[28] Amy K. Glasmeier, "New Data Up: Calculation of the Living Wage," MIT Living Wage Calculator, January 26, 2018, https://forr.com/irqn2_29.

[29] Irene Tung, Yannel Lathrop, and Paul Sonn, "The Growing Movement for $15," National Employment Law Project, November 2015, https://forr.com/irqn2_30.

[30] Tim O'Reilly argues that automation will fuel short-sighted shareholder interests. Tim O'Reilly, *WTF?: What's the Future and Why It's Up to Us* (HarperBusiness, 2017).

[31] Anita Balakrishnan and Berkeley Lovelace Jr., "IBM CEO: Jobs of the future won't be blue or white collar, they'll be 'new collar,'" CNBC, January 19, 2017, https://forr.com/irqn2_32. This comment by IBM CEO Ginni Rometty at the 2017 annual World Economic Forum in Davos, Switzerland, is typical.

Chapter 3

[32] Erik Lacitis, "60 years ago: The famous Boeing 707 prototype barrel roll over Lake Washington," *The Seattle Times*, August 7, 2015, https://forr.com/irqn3_33.

[33] Tim Iacono, "General Motors: A Brief History," Seeking Alpha, May 28, 2009, https://forr.com/irqn3_34.

[34] Vint Cerf, "A Brief History of the Internet & Related Networks," Internet Society, https://forr.com/irqn3_35.

[35] "On the Origins of Google," National Science Foundation, August 17, 2004, https://forr.com/irqn3_36.

[36] Pierre Bienaimé, "This Chart Shows How The US Military Is Responsible For Almost All The Technology In Your iPhone," *Business Insider*, October 29, 2014, https://forr.com/irqn3_37.

[37] Anaele Pelisson and Dave Smith, "These tech companies make the most revenue per employee," *Business Insider*, September 6, 2017, https://forr.com/irqn3_38. Revenue per employee for these three companies was taken from a chart in this article.

[38] Deadmalls.com, https://forr.com/irqn3_39. Deadmalls.com is an independent, not-for-profit website best known for featuring shopping malls in the US that have failed or are in the process of failing.

[39] Dr. Mark van Rijmenam, "From Big Data to Big Mac; how McDonalds leverages Big Data," Datafloq, https://forr.com/irqn3_40.

[40] "Employee Scheduling Regulations," New York State Department of Labor, https://forr.com/irqn3_41. This is a description of the hearings and regulations in New York.

[41] Emily Price, "Pepper the Robot Has a New Job at HSBC Bank," *Fortune*, June 27, 2018, https://forr.com/irqn3_42.

[42] Alex Rosenblat, "When Your Boss Is an Algorithm," *The New York Times*, October 12, 2018, 2018, https://forr.com/irqn3_43.

[43] Craig Le Clair, "The Forrester Wave™: Robotic Process Automation, Q2 2018," Forrester report, August 6, 2018, https://forr.com/irqn3_44. Forrester writes extensively about the RPA market. UiPath, Automation Anywhere, and Blue Prism are Leaders in this evaluation.

[44] Steve Lohr, "'The Beginning of a Wave': A.I. Tiptoes Into the Workplace," *The New York Times*, August 5, 2018, https://forr.com/irqn3_45.

[45] IBM Think conference, March 2018, Las Vegas.

[46] Reviewed by Jake Frankenfield, "Smart Contracts," Investopedia, April 18, 2019, https://forr.com/irqn3_47. Investopedia defines smart contracts as "self-executing contracts with the terms of the agreement between buyer and seller being directly written into lines of code. The code and the agreements contained therein exist across a distributed, decentralized blockchain network."

[47] Katie Lobosco, "University of Phoenix is the latest college under investigation," *CNN Money*, July 29, 2015, https://forr.com/irqn3_48.

[48] Antoine Gara, "Ivory Tower In The Cloud: Inside 2U, The $4.7 Billion Startup That Brings Top Schools To Your Laptop," *Forbes*, September 25, 2018, https://forr.com/irqn3_49. Operating income shot up from a $1.4 million in the red to $5 million in the black.

Chapter 4

[49] "Labor Force Statistics from the Current Population Survey," US Bureau of Labor Statistics, https://forr.com/irqn4_50.

[50] "Security Guards and Gaming Surveillance Officers," US Bureau of Labor Statistics, https://forr.com/irqn4_51.

[51] Based on an interview with Olek via email, between March and December 2018.

[52] "Japan's Diet votes yes to more foreign care workers," *Nikkei Asian Review*, November 19, 2016, https://forr.com/irqn4_53. This article comments on the healthcare worker shortage in Japan.

[53] Malcolm Foster, "Aging Japan: Robots may have role in future of elder care," Reuters, March 27, 2018, https://forr.com/irqn4_54. This article describes SoftBank Robotics' Pepper healthcare robot initiative in Japan.

[54] "Japan to loosen strict immigration rules amid labour shortage," Al Jazeera, November 1, 2018, https://forr.com/irqn4_55. The government is reviewing policies to allow more blue-collar and other workers in the country as businesses, from hotels to construction, struggle to find staff.

[55] Pradeep, "Microsoft announces new Azure IoT spatial intelligence capabilities," Microsoft, June 5, 2018, https://forr.com/irqn4_56.

[56] Michele Pelino and Andrew Hewitt, "Case Study: Use IoT To Transform Your Office Into A Smart Building," Forrester report, February 1, 2018, https://forr.com/irqn4_57.

[57] "Industries at a Glance," US Bureau of Labor Statistics, https://forr.com/irqn4_58.

[58] "Discover the benefits of the Smart Grid," Pacific Gas and Electric, https://forr.com/irqn4_59.

[59] Stephen Lacey, "How Peer-to-Peer Energy Trading on the Blockchain Might Work," GTM, April 3, 2018, https://forr.com/irqn4_60.

[60] "Renewable energy's job market continues to grow — with no end in sight," Arcadia Power, https://forr.com/irqn4_61. Growth of renewable energy benefits more than just the environment; it also adds to jobs.

[61] "Navistar: Reducing Maintenance Costs up to 40 percent for Connected Vehicles," Cloudera, August 30, 2017, https://forr.com/irqn4_62.

[62] "Logisticians," US Bureau of Labor Statistics, https://forr.com/irqn4_63.

[63] "G-STAR RAW franchisee Denimwall sets course for future with real-time data analytics enabled by technology and services from Detego, Impinj and RIoT Insight," Impinj press release, January 8, 2016, https://forr.com/irqn4_64.

[64] Hanhaa, https://forr.com/irqn4_65.

[65] "Finistere Ventures & PitchBook Agtech Investment Review," PitchBook Data, September 25, 2017, https://forr.com/irqn4_66. The report covers key trends in agtech investment across multiple facets of the space, from indoor agriculture to plant science, along with perspectives gleaned from interviews with industry experts. Private investment in 2017: $1.3 billion and rising, with 102 unique investors in agtech venture capital in the past two years.

[66] Forrester interview with Anthony Atlas, VP of product development, CERES, November 13, 2018.

[67] Jeff Daniels, "Future of farming: Driverless tractors, ag robots," CNBC, September 16, 2016, https://forr.com/irqn4_69.

[68] Dan Charles, "Robots Are Trying To Pick Strawberries. So Far, They're Not Very Good At It," NPR, March 20, 2018, https://forr.com/irqn4_69.

[69] "Locomotive Firer: Career Info & Requirements," Study.com, https://forr.com/irqn4_70. The rate of employment growth for locomotive firers is expected to decline by 79% between 2016 and 2026 with the introduction of computerized systems. These jobs often involve challenging physical conditions.

[70] Barbara Bean-Mellinger, "Male vs. Female Statistics in the Workplace in America," June 28, 2018, https://forr.com/irqn4_71. For example, for our physical worker persona, we include occupations like agricultural (76.1% male); architectural and engineering (92.6% male); and construction (93.3% male).

[71] Nick Wingfield, "As Amazon Pushes Forward with Robots, Workers Find New Roles," *The New York Times*, September 10, 2017, https://forr.com/irqn4_72.

[72] Forrester interview with Sarcos, June 26, 2018.

Chapter 5

[73] Kai-Fu Lee, *AI Superpowers: China, Silicon Valley, and the New World Order* (Houghton Mifflin Harcourt, 2018).

[74] Jeff Immelt, Automation Anywhere Imagine Conference, New York, May, 2018. Jeff Immelt was the ninth chairman of GE and served as CEO for 16 years, transforming it into a simpler, stronger, and more focused digital industrial company.

[75] Binyamin Appelbaum, "Nobel in Economics Is Awarded to Richard Thaler," October 9, 2017, *The New York Times*, https://forr.com/irqn5_76.

[76] Noel Sharkey, "Alan Turing: The experiment that shaped artificial intelligence," BBC News, June 21, 2012, https://forr.com/irqn5_77.

[77] Mark Twain, *The Wit And Wisdom of Mark Twain*, ed. Paul Negri, (Chartwell Books, April 26, 2016). "The difference between the almost right word and the right word is really a large matter— 'tis the difference between the lightning-bug and the lightning."

[78] Forrester's base data and approach used Frey/Osborne data sets, O*NET data sets, and US Bureau of Labor Statistics job numbers for 800-plus occupations. Using this information, we mapped all occupations to 12 generic personas; estimated the automation potential for each persona; created timelines for progression and disruption; and showed estimates for deficits, dividends, and evacuees.

Chapter 6

[79] "Labor Force Statistics from the Current Population Survey," US Bureau of Labor Statistics, January 18, 2019, https://forr.com/irqn6_80.

[80] The "fingers on the keys" task, to be outsourced in the first place, had to be well structured, repetitive, documented, and a noncritical business function. This profile makes it easy for a software robot to do the task, and there are strong economics behind it.

[81] Ananya Bhattacharya, "56,000 layoffs and counting: India's IT bloodbath this year may just be the start," *Quartz India*, December 26, 2017, https://forr.com/irqn6_82.

[82] We compiled data to show the average revenue growth from 2001 to 2018 from these top five Indian BPO companies: Cognizant, HCL, Infosys, TCS, and Wipro. We collected this data from financial annual reports. "Cognizant Reports Fourth Quarter And Full Year 2018 Results," Cognizant press release, February 6, 2019, https://forr.com/irqn6_83; "Q3 FY '19 Results," HCL, https://forr.com/6_83_2; "Financials & Filings," Infosys, https://forr.com/irqn6_83_3; Tata Consultancy Services, https://www.tcs.com/investor-relations; and "Annual Report 2017-2018," Wipro, https://forr.com/irqn6_83_4.

[83] US Bureau of Labor Statistics, "Loan Officers," April 12, 2019, https://forr.com/irqn6_84. Estimates on both loan officers and underwriters are from the US Bureau of Labor Statistics, 2016.

[84] Forrester interview with Sean Naismith, general manager, Enova Decisions, July 2, 2018.

[85] HPA, a Cognizant company, "Building a Workforce of the Future," Case Study, April 2019, https://forr.com/irqn6_86.

[86] Forrester interview, January 2018. Enzo quit that machine shop and drove a cab for almost a year, then was hired by another machine shop.

[87] Gwynn Guilford, "The 100-year capitalist experiment that keeps Appalachia poor, sick, and stuck on coal," *Quartz*, December 30, 2017, https://forr.com/irqn6_88.

[88] "Kentucky Office of Drug Control Policy Releases 2017 Overdose Fatality Report," Commonwealth of Kentucky Justice & Public Safety Cabinet, July 25, 2018, https://forr.com/irqn6_89.

[89] Matt O'Connor, "Chest x-ray AI similar, but quicker than radiologists at detecting diseases," HealthImaging, November 21, 2018, https://forr.com/irqn6_90.

[90] Prasad Akella, "Why Robots Won't Inherit the Plant, IndustryWeek, May 15, 2018, https://forr.com/irqn6_91. The International Federation of Robotics projects the global robot population to increase from 1.8 million in 2016 to 3.0 million by 2020. Combining those two numbers, the 1.2 million new robots coming online should displace 6.7 million jobs, a large-sounding number. But context is important. There are 340 million global temporary and full-time manufacturing workers.

[91] Prasad Akella, "Why Robots Won't Inherit the Plant, IndustryWeek, May 15, 2018, https://forr.com/irqn6_92.
[92] Forrester interview with Prasad Akella, founder and CEO, Drishti, August 2018. Drishti has a focus on human-driven processes — a unique application of computer vision and deep learning in manufacturing.

[93] Hal Sirkin, Michael Zinser, and Justin Rose, "The Robotics Revolution: The Next Great Leap in Manufacturing," Boston Consulting Group, September 23, 2015, https://forr.com/irqn6_94.

Chapter 7

[94] Kathleen Elkins, "Multiple jobs, sleeping in cars and other ways Facebook workers struggle to get by," CNBC, September 26, 2017, https://forr.com/irqn7_95.

[95] Will Markow, Soumya Braganza, Bledi Taska, Steven M. Miller, and Debbie Hughes, "The Quant Crunch," Burning Glass Technologies, 2017, https://forr.com/irqn7_96. Burning Glass Technologies analyzed 26.9 million US job postings from 2015. Postings for analytics skills in 2015 were 2.3 million. The forecast for 2020 was 2.7 million.

[96] Craig Le Clair, "RPA Operating Models Should Be Light And Federated," Forrester report, August 18, 2017, https://forr.com/irqn7_97. In this report, Forrester interviewed companies that implemented RPA at scale in their operations. The number of new automation jobs being added was averaged into a general formula and published.

[97] Dividends and deficits are shown as a percentage of today's employment. Automation dividends will be 13% of today's US employment of 155.8 million jobs, or 20.2 million new jobs created.

[98] Ajay Agrawal, Joshua Gans, and Avi Goldfarb, *Prediction Machines: The Simple Economics of Artificial Intelligence* (Harvard Business Review Press, 2018).

[99] "Revenue of the yoga industry in the United States from 2012 to 2020 (in billion U.S. dollars)," Statista, https://forr.com/irqn7_100.

[100] "The U.S. Meditation Market is Experiencing Strong Growth, With 9.3 Million Americans Meditating," *Business Wire*, May 2, 2018, https://forr.com/irqn7_101.

[101] "More Than 150 Million Americans Play Video Games," Entertainment Software Association press release, April 14, 2015 https://forr.com/irqn7_103.

[102] "A Global Community of Leaders," B Lab, https://forr.com/irqn7_104. B corporation certification is issued to for-profit companies. Companies must take an online assessment for social and environmental performance and receive a minimum score to be granted this certification.

[103] Richard Fry, "Millennials are the largest generation in the U.S. labor force," Pew Research Center, April 11, 2018, https://forr.com/irqn7_105.

[104] Masahiro Morioka, "A Phenomonelogical Study of 'Herbivore Men'" *The Review of Life Studies*, Vol. 4, September 2013, https://forr.com/irqn7_106. The rise of the "herbivore" men is often attributed to the advancement of women in society in the 1980s and the downturn of the Japanese economy in the 1990s.

[105] "Part-time employment rate," Organisation for Economic Co-operation and Development, https://forr.com/7_107.

[106] Kosugi Reiko, "Youth Employment in Japan's Economic Recovery: 'Freeters' and 'NEETs,'" The Asia-Pacific Journal, May 6, 2006, https://forr.com/irqn7_108.

[107] Marie Kondo, *The Life-Changing Magic of Tidying Up: The Japanese Art of Decluttering and Organizing* (Ten Speed Press, 2014). Marie Kondo points to a rise of minimalism in this lifestyle guide.

Chapter 8

[108] Based on an interview with ALTEN Calsoft Labs on April 13, 2018.

[109] Forrester surveyed 300 employees at director level and above, from manufacturing and high tech; banking and insurance; retail and CPGs; telecom and CSPs; and healthcare and life sciences, with minimum annual revenue of $500 million. This was a commissioned study conducted on behalf of Infosys' Edgeverve software division in June 2018.

[110] Ashley Hamer, "Moravec's Paradox Is Why the Easy Stuff Is Hardest for Artificial Intelligence," Curiosity.com, June 11, 2018, https://forr.com/irqn8_112.

[111] "Global Cobots Market Could be Worth $9 Billion by 2025," *Assembly*, September 4, 2018, https://forr.com/irqn8_113.

[112] Based on an interview with Sarcos on June 26, 2018. The company plans to offer three exoskeletons over the next few years, each giving wearers a different degree of strength and endurance support. For example, the Guardian XO will let wearers lift 80 pounds, while the Guardian GT —with 7-foot arms — will handle upward of 10,000 pounds.

[113] Nathaniel Popper, "The Robots Are Coming for Wall Street," *The New York Times Magazine*, February 26, 2016, https://forr.com/irqn8_115.

[114] Based on an interview with Cognizant in 2018.

[115] Craig Le Clair and J.P. Gownder, "The Future Of White-Collar Work: Sharing Your Cubicle With Robots," Forrester report, June 22, 2016, https://forr.com/irqn8_117. We interviewed IKEA as part of the research for this Forrester report. The interview was arranged by NICE, the provider of the RPA platform used for this automation.

[116] "The Future of Jobs Report 2018" World Economic Forum, September 17, 2018 https://forr.com/irqn8_118. According to the key findings of the report, of these employees, about "35% are expected to require additional training of up to six months, 9% will require reskilling lasting six to 12 months, while 10% will require additional skills training of more than a year."

[117] "Robot Bartenders Shake Things Up At Sea," Royal Caribbean, September 20, 2016, https://forr.com/irqn8_119. Created by Makr Shakr, an Italian robotics company based in Turin, the technology has three components: the robots, the core of the show that everyone can see; the tablet application that drives them; and the aesthetics or human characteristics.

[118] Based on an interview with MSC Cruises on July 2, 2018.

[119] Based on a briefing with Jim Walker, former shared services portfolio manager at NASA, and Christine Gex, former chief information security officer at the NASA Shared Services Center, on May 23, 2017.

[120] "America's #1 Health Problem," The American Institute of Stress, https://forr.com/irqn8_122.

[121] Jeffrey Pfeffer, *Dying for a Paycheck: How Modern Management Harms Employee Health and Company Performance — and What We Can Do About It* (HarperBusiness, 2018).

[122] David K. Johnson and Samuel Stern, "Focus On Employees' Daily Journeys To Improve Employee Experience" Forrester report, April 20, 2018, https://forr.com/irqn8_124. A traditional view of employee experience (EX) spans the employment life cycle from recruitment to exit and is too broad to uncover and address all the issues that most affect employees' daily work. Our research shows that daily work, and employees' ability to succeed with it, is more important. As a result, focusing on the technologies that enable employees to succeed in their daily work is critical.

[123] Aaron Smith and Monica Anderson, "Automation in Everyday Life," Pew Research Center, October 4, 2017, https://forr.com/irqn8_125.

Chapter 9

[124] "Costs of an Employee Vs. Independent Contractor," Chron, https://forr.com/irqn9_126.

[125] TJ McCue, "57 Million U.S. Workers Are Part Of The Gig Economy," *Forbes*, August 31, 2018, https://forr.com/irqn9_127.

[126] James Manyika, Susan Lund, Jacques Bughin, Kelsey Robinson, Jan Mischke, and Deepa Mahajan, "Independent work: Choice, necessity, and the gig economy," MicKinsey & Company, October 2016, https://forr.com/irqn9_128.

[127] Matthew Frankel, "9 Baby-Boomer Statistics That Will Blow You Away," The Motley Fool, July 29, 2017, https://forr.com/irqn9_129.

[128] US Bureau Of Labor Statistics, "Customer Service Representatives," https://forr.com/irqn9_130.

[129] "8 ways modern call center agents improve customer experience," liveops, https://forr.com/irqn9_131.

[130] Judith Solomon, "Medicaid Work Requirements Can't Be Fixed," Center on Budget and Policy Priorities, January 10, 2019, https://forr.com/irqn9_132.

[131] "Recognizing Africa's Informal Sector," African Development Bank Group, March 7, 2013, https://forr.com/irqn9_133.

Chapter 10

[132] Peggy Noonan, "A Life's Lesson," *The Time of Our Lives: Collected Writings*, Twelve, 2015.

[133] Jia Wertz, "Taking Risks Can Benefit Your Brand — Nike's Kaepernick Campaign Is A Perfect Example," *Forbes*, September 30, 2018, https://forr.com/irqn10_134.

[134] Paul Fain, "Latest Data on Enrollment Declines," *Inside Higher Ed*, May 29, 2018, https://forr.com/irqn10_136.

[135] Based on comments from Becky Schmitt, senior VP of global people at Walmart, at the MIT Future of Work Conference, November 2018.

[136] "Walmart Academy and Mobile Learning at BoxWorks 2017," Box video, https://forr.com/irqn10_138. "There was one woman with a lot of family and friends who had joined the celebration. Per Candace Davis, director of content at Walmart Academy: 'She said, 'I never graduated from high school. This is my first opportunity to graduate from anything, and for my children to see me be successful at anything.''"

[137] Leo Lewis, "Turning Off The Lights Is No Fix For Overworked Japan," *Financial Times*, October 18, 2016, https://forr.com/irqn10_139.

[138] Protesters in France have adopted the high-visibility yellow vest as the symbol of their movement. These are required in vehicles in France as a safety measure, but the color also speaks to the origin of many protesters who work in construction and related work categories. "Yellow Vest movement: Hundreds Arrested," BBC video, December 8, 2018, https://forr.com/irqn10_140.

[139] "Clean Jobs America 2019," E2, March 13, 2019, https://forr.com/irqn10_141. This a review of job openings and levels in the clean-energy economy. Many job segments are growing at over 10% a year.

[140] Chris Gardner and Craig Le Clair, "Get Control Over Your Bots With Forrester's Automation Framework," Forrester report, March 12, 2019 https://forr.com/irqn10_142.

[141] Alan S. Blinder, "Keynesian Economics," The Library of Economics and Liberty, https://forr.com/irqn10_143.

[142] Gene Callahan, "What is an Externality?" Mises Institute, August 1, 2001, https://forr.com/irqn10_144. "British economist A.C. Pigou was instrumental in

developing the theory of externalities. The theory examines cases where some of the costs or benefits of activities 'spill over' onto third parties. When it is a cost that is imposed on third parties, it is called a negative externality."

[143] Kevin J. Delaney, "The robot that takes your job should pay taxes, says Bill Gates," *Quartz*, February 17, 2017, https://forr.com/irqn10_145. "Robots are taking human jobs. But Bill Gates believes that governments should tax companies' use of them, as a way to at least temporarily slow the spread of automation and to fund other types of employment."

[144] Adam Smith, *The Wealth of Nations*, (William Strahan, Thomas Cadell, 1776). *The Wealth of Nations* is the major work of the Scottish economist Adam Smith. The book offers the world's first collected descriptions of what builds nations' wealth and is today a fundamental work in classical economics. A primary concept is that market forces of supply and demand will work together as an "invisible hand" to stabilize and optimize our economic lives.

INDEX

2U, 28, 29, 114
5G, 12, 37
A.C. Pigou, 102, 120
Accenture, 52, 111
Adam Smith, 103, 121
    Invisible hand, 103, 121
Agriculture technology, 38, 115
Amazon, 4, 5, 6, 16, 17, 21, 31,
    39, 42, 53, 55, 82, 99, 116
Anxiety, 2, 13, 74, 79, 80, 97, 102
    Meditation, 66
    Stress, 66, 69, 79, 80, 97
Appalachia, 1, 6, 55, 56, 69, 70,
    102, 117
Apple
    iPhone, 4, 5, 21
Automation deficits, 2, 5, 6, 7, 10,
    20, 32, 47, 61, 65, 83, 88, 102
Automation dividends, 2, 6, 7,
    37, 41, 45, 51, 65
Baby Boomers, 67, 84
Bill Gates
    Microsoft, 1, 102, 121
    Robot tax, 102
Black box, 14, 93
Black mirror, 41
Blockchain, 26, 27, 36, 63, 65,
    114
Boeing
    707, 21
    737, 14
    Ethiopia, 14
    World War II, 21
Business process outsourcing
    (BPO), 52, 116
Cambridge Analytica, 17, 113
Centaur, 75, 76, 77, 78, 79
Ceres Imaging, 38, 39
Change management, 9, 42, 52,
    65

Chatbots, 5, 26, 36, 43, 86
Coal mines, 3, 56
Cobots, 61
Cognitive tipping point (CTP), 41
Cognizant, 52, 77, 116, 117, 119
Constructive ambition, 53, 55, 93
Convergence, 37, 39, 40
DARPA, 21
David White, 6, 67
Denimwall, 38, 115
Digital Keynesian, 102
Drishti, 60, 117
eCommerce, 42
Edge computing, 31, 37
Eli Whitney, 45
Elon Musk, 1
    Tesla, 12, 17, 111
Entertainment Software
    Association, 67, 118
Exoskeleton, 34, 73
Facebook, 16, 17, 18, 21, 53, 63,
    112, 113, 117
Forces of automation, 1, 2, 47,
    105
Forrester
    David Johnson, 79, 105
    Glenn O'Donnell, 100, 105
    J.P. Gownder, 44, 93, 105, 111,
        119
    Martha Bennett, 27
    Rich Lane, 6, 81
    Ted Schadler, 12, 105
Freedom fighters, 67
Future of work, 1, 2, 4, 5, 10, 20,
    28, 34, 56, 70, 95, 98, 105, 107,
    109
Gene Roddenberry, 43
General Electric (GE), 42
General-purpose technology, 2
Gig economy, 4, 6, 7, 20, 81, 82,

83, 89, 119
Goldman Sachs, 75
Google, 11, 15, 16, 17, 21, 42, 44,
    53, 111, 112, 113
    Alphabet, 11, 15, 21
    Eric Schmidt, 15
Governance, 9, 27, 78, 94
Grant Thornton, 74
HSBC
    Pepper, 25, 114
IBM
    Ginni Rometty, 20, 113
    Watson, 14, 111
IKEA, 77, 119
Internet of things (IoT), 2, 31,
    32, 34, 37, 38, 115
Japan, 31, 33, 34, 60, 67, 97, 114,
    118, 120
    Herbivore, 67, 118
Jobcase, 98
    Fred Godd, 98
Kai-Fu Lee, 41, 116
Kiva robots, 40
Machine learning, 2, 4, 5, 6, 9, 17,
    21, 22, 25, 37, 38, 44, 53, 63,
    65, 75, 77, 81, 92, 93
Mark Zuckerberg
    Facebook, 15, 17, 112, 113
    Richard J. Durbin, 17
Martha Bennett, 105
McDonald's, 22, 23
Microgrid, 36
Millennials, 43, 67, 118
Netflix, 4, 5, 16, 42
Nick Mullins, 6, 56, 59, 69
Nike, 95, 120
Oculus Go, 96
Online education, 25, 29, 30
Panasonic, 34
Pandr Design Co., 84, 85
    Phoebe Cornog, 84
    Roxy Prima, 6, 84
Peggy Noonan, 95, 120

Personas
    Coordinators, 5, 25, 37, 38, 52,
        86, 94
    Cross-domain knowledge
        workers, 60, 63, 74, 99
    Cubicle workers, 5, 38, 49
    Digital elites, 6, 18, 21, 63, 65,
        93, 102
    Function-specific knowledge
        workers, 53, 74, 93
    Human-touch workers, 33, 63,
        66, 99
    Location-based workers, 36,
        59
    Mission-based workers, 9, 63,
        67, 69
    Physical workers, 39, 60, 65,
        73, 92, 115
    Single-domain knowledge
        workers, 59, 66, 94
    Teachers and explainers, 28,
        29, 30, 60, 63, 66, 88
Pew Research Center, 80, 118,
    119
Precision agriculture, 38
Predictability, 14
    Unpredictability, 13
Pymetrics, 15, 112
Richard Thaler, 42, 116
Robotics quotient (RQ), 93, 94
Royal Caribbean Cruise Lines, 78
S-curve, 41, 42, 44
Sebastian Zeiss, 71
Self-driving cars, 11, 13, 65
Software robots, 49, 53, 72, 91
Spatial intelligence, 34, 115
Splunk, 81
Stephen Hawkins, 1
Supply chain, 37, 55
Talent economy, 2, 7, 9, 28, 79,
    83, 84, 85, 88, 89, 97, 98, 100,
    101
Tata Consulting Group, 52, 116

Telecommunications, 75
Texas Children's Hospital, 77
Thelonius Monk, 13
Toronto moments, 14
Uber, 4, 22, 25, 38, 82
US Bureau of Labor Statistics, 7,

32, 109, 114, 115, 116
Virtual agents, 26, 76
Walmart, 96, 120
Wipro, 52, 116
World Economic, 20, 78, 113,
    119

Made in the USA
Middletown, DE
17 July 2019